Vladimir G. Bordo and Horst-Günter Rubahn

Optics and Spectroscopy
at Surfaces and Interfaces

Related Titles

M. Birkholz
Thin Film Analysis by X-Ray Scattering
approx. 400 pages with approx. 120 figures
Hardcover
ISBN 3-527-31052-5

W. Schmidt
Optical Spectroscopy in Chemistry and Life Sciences
An Introduction
384 pages with 269 figures and 14 tables
2005
Softcover
ISBN 3-527-29911-4

J. Berakdar, J. Kirschner (eds.)
Correlation Spectroscopy of Surfaces, Thin Films,
and Nanostructures
255 pages with 132 figures
2004
Hardcover
ISBN 3-527-40477-5

H.-J. Butt, K. Graf, M. Kappl
Physics and Chemistry of Interfaces
373 pages with 194 figures and 20 tables
2003
Softcover
ISBN 3-527-40413-9

R. J. H. Clark, R. E. Hester
Spectroscopy for Surface Science
Advances in Spectroscopy. Volume 26
420 pages
1998
Hardcover
ISBN 0-471-97423-4

Vladimir G. Bordo and Horst-Günter Rubahn

Optics and Spectroscopy
at Surfaces and Interfaces

WILEY-VCH

WILEY-VCH Verlag GmbH & Co. KGaA

The Authors

Vladimir G. Bordo
A. M. Prokhorov General Physics Institute
Moscow
Russia
e-mail: bordo@space.ru

Horst-Günter Rubahn
University of Southern Denmark
Odense/Sonderborg
Denmark
e-mail: rubahn@fysik.sdu.dk

Cover picture
a, b: Brilliant sodium clusters, embedded in an organic film. Force microscopy.
c, d: Blue light emitting organic nanofibers on a surface. Fluorescence microscopy.
All images: H.-G.Rubahn, Odense.

All books published by Wiley-VCH are carefully produced. Nevertheless, authors, editors, and publisher do not warrant the information contained in these books, including this book, to be free of errors. Readers are advised to keep in mind that statements, data, illustrations, procedural details or other items may inadvertently be inaccurate.

Library of Congress Card No.: applied for.

British Library Cataloging-in-Publication Data:
A catalogue record for this book is available from the British Library.

Bibliographic information published by Die Deutsche Bibliothek
Die Deutsche Bibliothek lists this publication in the Deutsche Nationalbibliografie; detailed bibliographic data is available in the Internet at <http://dnb.ddb.de>.

© 2005 WILEY-VCH Verlag GmbH & Co. KGaA, Weinheim

Printed in the Federal Republic of Germany
Printed on acid-free paper

Typesetting   Uwe Krieg, Berlin
Printing   betz-druck GmbH, Darmstadt
Binding   J. Schäffer GmbH i.G., Grünstadt

ISBN-13: 978-3-527-40560-2
ISBN-10: 3-527-40560-7

Contents

Optics and Spectroscopy at Surfaces and Interfaces. Vladimir G. Bordo and Horst-Günter Rubahn
Copyright © 2005 WILEY-VCH Verlag GmbH & Co. KGaA, Weinheim
ISBN: 3-527-40560-7

Preface

Optical tools for surface and interface analysis have become very prominent
working horses within the last twenty years due to their steadily increasing
sensitivity, their sensitivity in probing and their inherent possibilities for *in
situ* operation. Recently organized conference series such as "Laser Techniques for Surface Science" and "Optics of Surfaces and Interfaces" have
reflected this development. Of course, this is followed by an urgent demand
for updated educational programs to prepare specialists in this rapidly pro-
gressing field.

This book is based partially on graduate courses held at the Physics De-
partment of the University of Southern Denmark in Odense. While there is
a large variety of excellent books specializing either in surface science or in
optics and spectroscopy, a self-consistent book on optics and spectroscopy of
surfaces and interfaces is apparently missing. It is this gap that our textbook
tries to fill at a level which is accessible for students and beginners in this
field but with enough depth to make it interesting also for the advanced user
of optical techniques at surfaces. We realize that optical interface analysis
has found applications in diverse fields of science and technology and has
become an essential tool in physics, chemistry, materials and life sciences. The
potential readers of this book will therefore have rather different knowledge
levels, but we hope that most of them will benefit from reading some parts or
the complete book.

One of us (VGB) would like to express warm gratitude to his wife, Elena,
for her patience and encouragement without which the writing of this book
would have been impossible.

Moscow and Odense, April 2005

Vladimir G. Bordo and Horst-Günter Rubahn

Optics and Spectroscopy at Surfaces and Interfaces. Vladimir G. Bordo and Horst-Günter Rubahn
Copyright © 2005 WILEY-VCH Verlag GmbH & Co. KGaA, Weinheim
ISBN: 3-527-40560-7

1
Introduction

Physics and chemistry of surfaces and interfaces is one of the most challenging areas of modern science, in view of its numerous applications in various fields of science and technology. Various technologically important processes occur either in the surface region of crystals or at the boundary between different media. For example, the fast development of microelectronics was promoted by a deepening of our understanding of electronic processes at interfaces between a semiconductor and a metal or another semiconductor. Heterogeneous catalysis is essentially chemistry at surfaces and thus demands a thorough knowledge of molecular processes at a gas–solid interface. Electrochemistry is based on similar processes that occur at an interface between a metal electrode and an electrolyte solution. Heat and mass transfer as well as the phenomenon of friction are determined by physical properties of an interface between contacting media as well as by elementary processes which occur on it. Such important modern techniques as molecular beam epitaxy and chemical vapor deposition are directly related to phenomena at solid surfaces. This list is far from being complete and can be continuously increased.

The main problem which an investigator faces in surface and interface science is that the number of atoms or molecules in the surface or interface region is small compared with those in the bulk. It is therefore necessary to have specific analytical tools at hand which are sensitive to surface and interface properties as well as to processes which occur in those regions. Among the different analytical techniques which have been developed in surface science over the years, optical techniques have definite advantages determined by their non-perturbing character, reliability and ease of use. They do not require ultrahigh vacuum and can be applied under good vacuum and ambient conditions as well, thus allowing one *in situ* investigations. To bring the whole into context we note that, in fact, this very branch of surface science dates back to 1890 when Drude studied the reflection of light from crystal surfaces in the presence of surface contamination. In this sense optical methods were the first to allow truly surface-specific investigations.

Optics and Spectroscopy at Surfaces and Interfaces. Vladimir G. Bordo and Horst-Günter Rubahn
Copyright © 2005 WILEY-VCH Verlag GmbH & Co. KGaA, Weinheim
ISBN: 3-527-40560-7

This book begins with an introductory chapter to surface and interface science. Different properties of crystal surfaces and their interfaces to liquids and gases are considered. The next chapter is devoted to linear optical properties of surfaces as well as of interfaces between two media. Although the contacting media are described in terms of their dielectric functions, the relevant theory is equally valid for solids, liquids and gases, except some explicitly specified cases. Analytical techniques considered in this book can be broadly classified as infrared and optical where the term *optical* implies the use of electromagnetic radiation in the visible spectral range or close to it. With the invention of lasers operating in different spectral regimes, the possibilities of optical analytical techniques have been extended considerably. Both linear and nonlinear laser spectroscopy is widely used nowadays in surface and interface analysis and thus is discussed with some broadness in this book. Lasers provide high brightness, monochromaticity and directivity of the probing light. Ultrashort pulsed lasers allow one to monitor temporal behavior of surface processes with unprecedented precision.

An interface between two media is accessible for light if at least one of the media is transparent in the considered spectral region. Therefore, optical tools can be exploited for interface analysis in the ranges of transparency of the contacting media. However, the range of applicability of optical techniques can be significantly extended if the incident light excites electromagnetic modes bound to the interface. Such electromagnetic waves are called surface polaritons. For example, at a metal surface they can be excited in a wide spectral range from the far infrared to the far ultraviolet region. The use of surface polaritons for surface and interface analysis is therefore discussed at various places in this book.

A measurement of the reflection coefficient of light is the most simple and oldest optical technique, providing information on a surface or an interface. Being applied to an interface between a solid and a gaseous phase it reflects the features of the gas optical spectrum. However, it was only discovered during the fifties of the last century, that at relatively low gas pressures the reflection of light resonant to the transition in gas atoms in such a system displays peculiarities which can be attributed to gas–surface scattering. This relatively new field of optical spectroscopy has received growing attention recently, prompting us to discuss it in some more detail.

A traditional branch of surface optics is optical microscopy, i.e., imaging of objects disposed on a surface. Since the invention of the first optical microscopes in the 16th century this field has undergone strong and permanent progress. In addition to bright-field, microcopy images can also be obtained by dark and near-field microscopy, resulting in large contrast improvement and better and better resolution. Sophisticated implementation of laser scanning techniques has led to what one usually calls ultramicroscopy. Easy access to three-dimensional images becomes possible via confocal approaches. Finally, the use of nonlinear optical and polarization techniques gives access even to buried interfaces.

We conclude the book with a discussion of optical tools with high spatial resolution that match recent demands from nanoscience and nanotechnology. This new class of optical devices, such as scanning near-field optical microscopes or photon scanning tunneling microscopes, has been developed to investigate surface properties at the nanometer scale. The same tools are capable of registering optical spectra locally, reaching the limit of single molecule spectroscopy.

This book is a textbook and as such it can be used in a consecutive way for lecturing chapter by chapter. However, we have tried to make each chapter also understandable in itself, considerably restricting the number of cross references. Throughout the text, references to specific literature are given. The interested reader can find more general literature on the topics of the individual chapters at the end in a Further Reading section.

Some of the theoretical formulation is not very straightforward. Therefore, we have printed the text in small letters which can be skipped at a first short reading. We suggest working through the problems that are given at the end of the chapters which should should serve to strengthen the readers capability of handling problems of optics at surfaces. Solutions are added at the very end of the book.

Finally, in order to avoid multiple definitions of the same terms we have added a brief glossary and also an index that allows one to look up specific topics.

2
Surfaces and Interfaces

Let us begin with the definition of surfaces and interfaces in terms of the properties which we are intending to investigate. In general, an interface is that region containing the border surface where some properties (electronic, vibrational, optical, etc.) differ from those in the bulk of the two contacting media. In terms of its atomic structure, the surface region includes a few atomic layers where the geometrical arrangement of the atoms is distorted by surface forces. With respect to the electrical properties of semiconductor surfaces, we recognize that a rather extended layer, having a thickness of hundreds of Ångstroms, is relevant for the determination of surface properties. In the gas phase, the very dimensions of the interface layer (the so-called Knudsen layer) depend on the gas pressure. That region which we call the interface or surface region, therefore, should be determined separately in every case.

2.1
Solid Surfaces

Optical properties of surfaces are determined to a great extent by structure and morphology. The term "structure" implies the geometrical arrangement of atoms in the crystal lattice. It reflects the detailed microscopic picture of a surface under equilibrium conditions. In contrast, "morphology" characterizes the macroscopic shape of a surface. These two distinct features of surfaces manifest themselves under different conditions which are determined by the spatial resolution of the surface probe. When the surface is irradiated by X-rays whose wavelength is comparable with the interatomic distance in a crystal, the microscopic structure is probed. The same is true for fast particle beams. For example, He atoms with a kinetic energy of 20 meV have a de Broglie wavelength of 1 Å and hence represent a surface probe with microscopic resolution. For the optical spectral range the typical wavelengths are essentially longer, of the order of several hundred nanometers. Correspondingly, light sources probe the surface morphology rather than the surface structure.

Optics and Spectroscopy at Surfaces and Interfaces. Vladimir G. Bordo and Horst-Günter Rubahn
Copyright © 2005 WILEY-VCH Verlag GmbH & Co. KGaA, Weinheim
ISBN: 3-527-40560-7

Another critical point is the penetration depth of a surface probe into a crystal. X-rays and various kinds of particles (electrons, atoms, molecules, ions) are traditionally used for surface analysis. By choosing their energies or the angle of incidence of the beams, one can achieve a penetration depth in the nanometer range or even shorter, providing a high surface sensitivity. In the case of optical probes, the penetration depth is determined by the absorption coefficient. For absorption bands in solids this coefficient is of the order of 10^4–10^5 cm^{-1} which means that light penetrates into a solid over distances of 0.1–1 μm. A solid slab of such thickness contains more than one hundred atomic planes and the contribution of a few atomic layers nearby the surface is negligible. This example illustrates how crucial the question of surface sensitivity is in surface optics.

Surface atoms can be transferred into an excited state. Different types of surface excitations (vibrational or electronic) result in features within different spectral ranges (infrared and visible or around it, respectively) and they require different light sources and detectors for their investigation. It is reasonable, therefore, to classify the corresponding spectroscopic techniques as infrared and optical, where the term "optical" is used for electromagnetic radiation in and around the visible region of the spectrum.

2.1.1
Surface Structure

2.1.1.1 Relaxation and Reconstruction

An ideal crystal is characterized by a regular arrangement of its constituent atoms. Its microscopic structure possesses a three-dimensional (3D) translational invariance which means that a crystal can be constructed by an infinite repetition of a unit cell along three independent directions. A surface eliminates this invariance along one of the directions. It retains a symmetry with respect to two-dimensional (2D) translations along the surface itself.

Upon crystal cleavage the surface atomic layer, as well as a few adjacent layers, have a different local environment compared with the atomic layers deep inside the crystal. The forces from the atoms in the crystal interior acting on the surface atoms are no longer compensated by the removed part of the crystal. As a consequence, the net force applied to the surface atoms tends to rearrange their equilibrium positions. The possible types of surface rearrangements fall into two main classes. A modification of the distances between the atomic planes near the surface is called *relaxation*. It is not accompanied by changes in the topmost surface layer structure compared with the structure in the bulk. Under a surface *reconstruction*, however, the surface atoms are shifted along the surface with respect to their positions in the bulk. As a result, the surface unit cell differs from the one which would exist if there were no displacements of surface atoms.

Which of these surface rearrangements takes place upon cleavage is determined by the surface electronic structure. This follows within the *adiabatic approximation* which is applicable for a wide range of molecular systems and is based on the fact that the ratio m_e / M is small. Here, m_e and M are the electron and the mean nuclei masses, respectively. In the zeroth-order approximation, which corresponds to infinite masses of nuclei, one can neglect their kinetic energies and consider the states of the electronic subsystem at fixed nuclei positions specified by a multidimensional vector $\mathbf{R}$. Then the electronic energy $E(\mathbf{R})$ plays the role of a potential in which the nuclei move. The equilibrium positions of the nuclei, $\mathbf{R}_0$, are found at the minimum

$$\left(\frac{\partial E(\mathbf{R})}{\partial \mathbf{R}} \right)_{\mathbf{R}=\mathbf{R}_0} = 0 \tag{2.1}$$

In metals, the electron density behaves somewhat like a fluid and tends to smooth out the surface relief created on cleavage. As a result, electronic surface dipoles appear in the surface region which cause inward displacements of surface atomic cores, i.e., relaxation. A qualitatively different case is found at semiconductor surfaces. A terminated semiconductor crystal contains on its surface so-called dangling bonds which originate from hybrid sp^3-orbitals of surface atoms. Upon their saturation the creation of new bonds between adjacent surface atoms leads to a lowering of the free surface energy. Such a rearrangement causes shifts of the atoms in the surface plane, i.e., a reconstruction.

2.1.1.2 Surface Lattice and Superstructure

In order to describe the positions of surface atoms, a 2D reference frame has to be introduced. Due to the 2D periodicity along the surface it is sufficient to specify the arrangement of atoms in a unit cell. Besides the translational invariance, a crystal possesses a symmetry with respect to rotations and reflections in mirror planes. At a crystal surface, these point-group operations have to leave the surface atoms in the surface plane. Therefore, only rotation axes and mirror planes perpendicular to the surface can be included in a surface symmetry group. The possible point-group operations must be compatible with 2D translations. This condition leads us to the following ten allowed point-group symmetries:

$$1, 1m, 2, 2mm, 3, 3m, 4, 4mm, 6, 6mm \tag{2.2}$$

Here, the digits $n = 1...6$ denote the axes of rotation, C_n, by an angle $2\pi/n$ and the letter m implies a mirror plane σ_v containing a rotation axis.

Any crystal plane can be specified by a 2D *Bravais lattice* constructed of primitive (p) 2D unit cells (parallelograms) with a lattice point at each corner. The symmetry considerations given above restrict the number of different

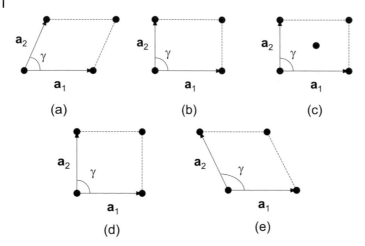

Fig. 2.1 Five surface Bravais lattices: (a) oblique (*p2*), $a_1 \neq a_2$, $\gamma \neq 90°$; (b) primitive rectangular (*p2mm*), $a_1 \neq a_2$, $\gamma = 90°$; (c) centered rectangular (*c2mm*), $a_1 \neq a_2$, $\gamma = 90°$; (d) square (*p4mm*), $a_1 = a_2$, $\gamma = 90°$; (e) hexagonal (*p6mm*), $a_1 = a_2$, $\gamma = 120°$.

possible 2D Bravais lattices to the five shown in Fig. 2.1. The centered rectangular unit cell is not primitive as it contains two lattice points (the other ones are shared by adjacent cells). Nevertheless, sometimes it is more convenient to use it for description. The primitive lattice vectors $\mathbf{a}_1$ and $\mathbf{a}_2$ generate a Bravais lattice

$$\mathbf{a_n} = n_1\mathbf{a}_1 + n_2\mathbf{a}_2 \tag{2.3}$$

where $\mathbf{n} = (n_1, n_2)$ and n_1 and n_2 are integers. Such a lattice describes the ideal crystal structure of a surface as if there were no rearrangements of surface atoms. However, due to reconstruction the actual surface structure may differ from that given by Eq. (2.3). Then, the new Bravais lattice

$$\mathbf{b_m} = m_1\mathbf{b}_1 + m_2\mathbf{b}_2 \tag{2.4}$$

with $\mathbf{m} = (m_1, m_2)$, specified by the primitive vectors $\mathbf{b}_1$ and $\mathbf{b}_2$, is called a *superlattice*. The vectors $\mathbf{b}_i$ can be expanded in terms of the vectors $\mathbf{a}_j$ as

$$\mathbf{b}_1 = c_{11}\mathbf{a}_1 + c_{12}\mathbf{a}_2$$
$$\mathbf{b}_2 = c_{21}\mathbf{a}_1 + c_{22}\mathbf{a}_2 \tag{2.5}$$

or, in matrix form,

$$\begin{pmatrix} \mathbf{b}_1 \\ \mathbf{b}_2 \end{pmatrix} = \mathbf{C} \begin{pmatrix} \mathbf{a}_1 \\ \mathbf{a}_2 \end{pmatrix} \tag{2.6}$$

where $\mathbf{C}$ is a square matrix

$$\mathbf{C} = \begin{pmatrix} c_{11} & c_{12} \\ c_{21} & c_{22} \end{pmatrix} \tag{2.7}$$

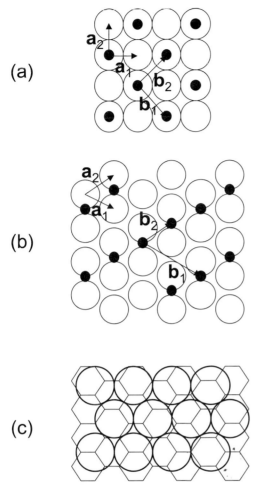

Fig. 2.2 Examples of different superlattices: (a) and (b) simple, (c) incoherent. In (a) and (b) the atoms of the superlattice and those of the substrate are shown by the filled and open circles, respectively. In (c) the large circles denote atoms adsorbed on the hexagonal substrate lattice.

The areas of the primitive cells of the two Bravais lattices are equal to $A = |\mathbf{a}_1 \times \mathbf{a}_2|$ and $B = |\mathbf{b}_1 \times \mathbf{b}_2|$. As follows from Eq. (2.6), they are related to each other by the equation

$$B = A \det \mathbf{C} \tag{2.8}$$

The value of $\det \mathbf{C}$ thus characterizes the distortion of the ideal crystal structure due to cleavage. When $\det \mathbf{C}$ is an integer, the superlattice is referred to as *simple* (Figs 2.2a, b). When $\det \mathbf{C}$ is a rational number, the superstructure is called *coincidence lattice*. In such a case, if there is no angle between the

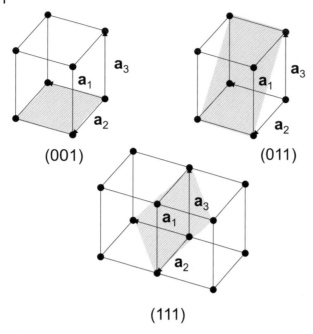

(001)   (011)

(111)

Fig. 2.3 Different faces of a cubic crystal. a_1, a_2 and a_3 are the primitive vectors of a 3D lattice. The denoted faces are shown shaded.

two Bravais lattices, the actual positions of surface atoms coincide over a few periods with their positions in the ideal surface structure.

The concept of a superlattice can also be applied to regular overlayers of foreign adsorbed atoms (see Section 2.2). If the coupling forces between atoms exceed the atom–surface interaction forces, they can form a structure which is not related to the symmetry of the substrate surface (*incommensurate structure* or *incoherent lattice*). In such a case det $\mathbf{C}$ is an irrational number (Fig. 2.2c).

2.1.1.3 **Surface Structure Notation**

To specify the surface structure it is necessary to give the crystal face under consideration a notation. A face is completely determined by its interceptions along the crystallographic axes in terms of the 3D unit cell dimensions. Let us take the reciprocals of those quantities, clear the fractions and reduce them to their lowest terms. Then one obtains three integers *{hkl}* which are called the *Miller indices*. Following convention, the face of a crystal X is given by *X(hkl)*. Negative indices are denoted by a bar above them. Figure 2.3 shows examples of notations for different faces of a cubic crystal. The Miller indices can also be used to identify the direction perpendicular to the crystal plane. In such a case the indices denoting the plane are set into square brackets: *[hkl]*.

If a reconstruction is described by the relations

$$\mathbf{b}_1 = p\mathbf{a}_1, \quad \mathbf{b}_2 = q\mathbf{a}_2 \tag{2.9}$$

then it is denoted as X(hkl) p×q. A superlattice with a centered unit cell is marked by the letter c: X(hkl) cp×q. If the vectors of the superstructure unit cell make an angle φ with those of the ideal structure, this is marked as X(hkl) p×q − φ. The structure of the adsorbate A overlayer is described by similar notations. In this case the adsorbate superstructure is set into brackets: X(hkl) (p×q − φ) − A. For example, the superlattices on surfaces of a simple cubic crystal represented in Figs 2.2a and b are denoted as X(100) ($\sqrt{2} \times \sqrt{2} - 45°$) − A and X(111) (2 × 1) − A, respectively.

2.1.1.4 Reciprocal Lattice

Besides the 2D Bravais lattice (2.3) in radius vector space, it is worthwhile to introduce a lattice in the space of wave vectors parallel to the surface. A basis in such a space can be determined as

$$\mathbf{g}_1 = 2\pi \frac{\mathbf{a}_2 \times \hat{\mathbf{n}}}{|\mathbf{a}_1 \times \mathbf{a}_2|}, \quad \mathbf{g}_2 = 2\pi \frac{\hat{\mathbf{n}} \times \mathbf{a}_1}{|\mathbf{a}_1 \times \mathbf{a}_2|} \tag{2.10}$$

where

$$\hat{\mathbf{n}} = \frac{\mathbf{a}_1 \times \mathbf{a}_2}{|\mathbf{a}_1 \times \mathbf{a}_2|} \tag{2.11}$$

is a unit vector along the surface normal. It can be readily proven that

$$\mathbf{g}_i \cdot \mathbf{a}_j = 2\pi\delta_{ij} \tag{2.12}$$

which means that the vectors $\mathbf{g}_i$ are orthogonal to the primitive vectors of the Bravais lattice and their lengths are reciprocal to those of the $\mathbf{a}_j$'s modulo a factor $2\pi/\sin\gamma$ with γ the angle between the vectors $\mathbf{a}_1$ and $\mathbf{a}_2$. The corresponding lattice generated by the vectors $\mathbf{g}_1$ and $\mathbf{g}_2$ is called a 2D *reciprocal lattice*.

The reciprocal lattice concept can be illustrated by constructing a Fourier series in the space of 2D periodic functions. Let $f(\mathbf{r})$ be a periodic function of a radius-vector in the surface plane, i.e.,

$$f(\mathbf{r} + \mathbf{a_n}) = f(\mathbf{r}) \tag{2.13}$$

with the vector $\mathbf{a_n}$ specified by Eq. (2.3) for any $\mathbf{n}$. Its Fourier series can be written in a general form as

$$f(\mathbf{r}) = \sum_{\mathbf{G}} f_{\mathbf{G}} \cdot \exp(i\mathbf{G} \cdot \mathbf{r}) \tag{2.14}$$

On the other hand, the periodicity of $f(\mathbf{r})$ implies that the equation

$$\exp[i\mathbf{G} \cdot (\mathbf{r} + \mathbf{a_n})] = \exp(i\mathbf{G} \cdot \mathbf{r}) \tag{2.15}$$

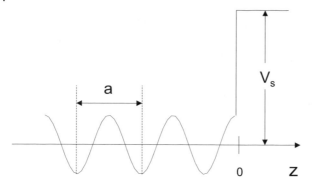

Fig. 2.4 The 1D one-electron potential. The semiaxis $z < 0$ corresponds to the crystal.

or, equivalently,

$$\exp(i\mathbf{G} \cdot \mathbf{a_n}) = 1 \tag{2.16}$$

must be fulfilled. The solutions of the latter equation obey the relation $\mathbf{G} \cdot \mathbf{a_n} = 2\pi m$, where m is an arbitrary integer. Therefore, taking into account Eq. (2.12) we conclude that the vector $\mathbf{G}$ must be represented in the form

$$\mathbf{G} = h_1 \mathbf{g_1} + h_2 \mathbf{g_2} \tag{2.17}$$

with h_1 and h_2 integers, i.e., it belongs to the reciprocal lattice.

2.1.2
Electronic Surface States

In the adiabatic approximation the states of the electronic subsystem are calculated at fixed positions of nuclei which are considered as parameters. That way the arrangement of nuclei is reflected in the Schrödinger equation for electronic states. Near the surface, the electronic states should thus be modified compared with those in the bulk crystal. To find them, one has to take into account the surface crystal structure in the equation for the electronic motion, which is in general a formidable problem. As in the case of bulk electronic states, a one-dimensional (1D) model is a useful first approach. Tamm (1932) was the first to demonstrate that the presence of a crystal surface itself leads to the existence of electronic *surface states* which are inaccessible to electrons in infinite crystals.

2.1.2.1 **One-dimensional Problem**
Let us consider a semi-infinite 1D chain of atomic cores along the negative z-semiaxis. Let their separation be a. We shall assume that the one-electron[1]

---

1) The term "one-electron" here means that each of the electrons moves in the field of the positive ions and in the average field of all the other electrons.

potential $V(z)$ is a periodic function of z of period a within the chain (Fig. 2.4). The end of the chain (a 1D surface) is represented by a potential barrier of constant height V_s. An electron wave function $\psi(z)$ in such a system obeys the Schrödinger equation

$$\left[-\frac{\hbar^2}{2m_e} \frac{d^2}{dz^2} + V(z) \right] \psi(z) = E\psi(z) \tag{2.18}$$

with m_e the mass of the electron. According to the Bloch–Floquet theorem, Eq. (2.18) with a periodic potential (i.e., within the chain) has the solutions

$$\psi_k(z) = u_k(z)e^{ikz} \tag{2.19}$$

where $u_k(z)$ is a periodic function of period a and the real quantity k which is called a *wave number* should be determined from the boundary conditions imposed on $\psi_k(z)$. Note that the parameter k is not uniquely determined by Eq. (2.19). One can construct a new wave number

$$\tilde{k} = k + \frac{2\pi n}{a} \tag{2.20}$$

and a new periodic function

$$\tilde{u}_{\tilde{k}}(z) = u_k(z)e^{-i(2\pi n/a)z} \tag{2.21}$$

where n is an integer. The new total wave function

$$\tilde{\psi}_{\tilde{k}}(z) = \tilde{u}_{\tilde{k}}(z)e^{i\tilde{k}z} \tag{2.22}$$

then is identical with the function $\psi_k(z)$ given by Eq. (2.19). Therefore, any solution of the form (2.19) can be reduced to an *equivalent* solution corresponding to the wave number k within the range $[-\pi/a, +\pi/a]$ which is called the *first Brillouin zone*.

2.1.2.2 Nearly-free Electron Approximation
In general, Eq. (2.18) cannot be solved exactly. Let us now discuss the *nearly-free electron approximation* where the potential $V(z)$ can be treated as a small perturbation to the motion of a free electron.

Solution within a chain
Since the functions $V(z)$ and $u_k(z)$ are periodic, they can be expanded in a Fourier series

$$V(z) = \sum_m V_m e^{i(2\pi m/a)z} \tag{2.23}$$

$$u_k(z) = \sum_n c_n e^{i(2\pi n/a)z} \tag{2.24}$$

where, without loss of generality, one can set $V_0 = 0$ and since the potential is real one has

$$V_{-m} = V_m^*$$ (2.25)

Substituting the expansion (2.24) into Eq. (2.19), we obtain

$$\psi_k(z) = \sum_n c_n \psi_{kn}^{(0)}(z)$$ (2.26)

where we have introduced the notation

$$\psi_{kn}^{(0)}(z) = e^{i(2\pi n/a)z} e^{ikz}$$ (2.27)

The functions (2.27) are the eigenfunctions of the unperturbed Hamiltonian $-(\hbar^2/2m_e)(d^2/dz^2)$ corresponding to the eigenvalues

$$E_{kn}^{(0)} = \frac{\hbar^2}{2m_e}\left(k + \frac{2\pi n}{a}\right)^2$$ (2.28)

The first-order correction to the energy is found from perturbation theory as

$$E_{kn}^{(1)} = \frac{\langle \psi_{kn}^{(0)}|V|\psi_{kn}^{(0)}\rangle}{\langle \psi_{kn}^{(0)}|\psi_{kn}^{(0)}\rangle} = 0$$ (2.29)

This can be shown as follows. To calculate the correction to the energies accounting for the influence of the potential $V(z)$, we first assume that the atomic chain is extended up to $-L$ ($L \gg a$) from the side of negative z and then tend L to infinity. Direct evaluation gives the matrix elements

$$\langle \psi_{kn}^{(0)}|\psi_{km}^{(0)}\rangle = L\delta_{mn} + \frac{\sin\{[\pi(m-n)/a]L\}}{[\pi(m-n)/a]}e^{-i[\pi(m-n)/a]L}(1 - \delta_{mn})$$ (2.30)

and

$$\langle \psi_{kn}^{(0)}|V|\psi_{km}^{(0)}\rangle = V_{n-m}L + \sum_{l\neq n-m}\frac{\sin\{[\pi(l+m-n)/a]L\}}{[\pi(l+m-n)/a]}V_l e^{-i[\pi(l+m-n)/a]L}$$ (2.31)

For $L \to \infty$, taking into account that

$$\lim_{L\to\infty}\frac{\sin[(\pi n/a)L]}{(\pi n/a)} = \pi\delta(\pi n/a)$$ (2.32)

we obtain

$$E_{kn}^{(1)} = V_0 = 0$$ (2.33)

Thus in the linear approximation in $V(z)$ the electron energies coincide with their zeroth-order values (2.28).

This result is valid everywhere in k-space where the perturbation theory can be applied, i.e., the different eigenvalues $E_{km}^{(0)}$ do not get close to each other.

Let us consider the solution $E_{k0}^{(0)}$ which corresponds to the electron energy in free space. Degeneracy with the other eigenvalues,

$$E_{k0}^{(0)} = E_{kn}^{(0)} \tag{2.34}$$

results in roots of the form

$$k = \pm \frac{\pi n}{a}, \quad n = 1, 2, 3, \dots \tag{2.35}$$

i.e., such critical points coincide with the boundaries of the nth Brillouin zones. In the vicinities of these points a quasi-degenerate perturbation theory must be applied. The zeroth-order wave function should be chosen in the form

$$\psi_k^{(0)}(z) = A\psi_{k0}^{(0)} + B\psi_{kn}^{(0)} \tag{2.36}$$

where A and B are constants. Substituting (2.36) into the Schrödinger equation (2.18), we find in matrix form a system of equations for A and B

$$\begin{pmatrix} (\hbar^2 k^2/2m_e) - E_k & V_{-n} \\ V_n & (\hbar^2 \tilde{k}^2/2m_e) - E_k \end{pmatrix} \begin{pmatrix} A \\ B \end{pmatrix} = 0 \tag{2.37}$$

with $\tilde{k}$ defined in Eq. (2.20). This system of equations has a nontrivial solution only if its determinant equals zero. Thus we find from here

$$E_k^{\pm} = \frac{\hbar^2(k^2 + \tilde{k}^2)}{4m_e} \pm \sqrt{\left[\frac{\pi \hbar^2 n}{m_e a}\left(k + \frac{\pi n}{a}\right)\right]^2 + |V_n|^2} \tag{2.38}$$

where we have used the property (2.25). From Eq. (2.38) it follows that there are two energy branches $E_k^{\pm}$ which are split at the nth Brillouin zone boundary by $2|V_n|$. Therefore, the energy spectrum consists of *allowed energy bands* (AEB) separated by *forbidden energy gaps* (FEG), where no solutions exist for real k-values (Fig. 2.5).

Solution at the surface

Since the Hamiltonian in the Schrödinger equation (2.18) is real, both the wave functions $\psi_k(z)$ and $\psi_k^*(z)$ are eigenfunctions at the same energy E_k. Taking into account that

$$\psi_k^*(z) = \psi_{-k}(z) \tag{2.39}$$

one can write the general solution within the chain as

$$\psi_c(z) = \alpha u_k(z)e^{ikz} + \beta u_{-k}(z)e^{-ikz} \tag{2.40}$$

where α and β are constants. Inside an infinite crystal the quantity k must be real to ensure the finiteness of the wave function. However, at the surface this is not necessary. Instead, a wave number of the form

$$k = q + i\kappa \tag{2.41}$$

i.e., with an imaginary part $i\kappa$, is also allowed. The solution (2.40) along with its derivative must be matched at $z = 0$ with the wave function from the vacuum side ($z > 0$)

$$\psi_v(z) = \gamma e^{-k_0 z} \tag{2.42}$$

where

$$k_0 = \frac{1}{\hbar}\sqrt{2m_e(V_s - E)} \tag{2.43}$$

with $E < V_s$. When k is real, one obtains two linear equations with respect to α, β and γ which obviously can be solved for any values of the coefficients, i.e., for any E inside the allowed band. This means that all the solutions found for an infinite crystal exist at the surface, too. However, when k is a complex quantity, i.e., the corresponding energy falls into the forbidden gap, the number of unknown variables is reduced. Indeed, let us assume for definiteness that $\kappa > 0$. Then the wave function (2.40) is finite at $z \to -\infty$ if $\alpha = 0$. The remaining two coefficients, β and γ, can be found if the determinant of the system is equal to zero. This condition provides the energies of the corresponding states. In such a case, the electron wave function decays exponentially on both sides from the point $z = 0$, i.e., the electron in such a *surface state* is localized near the crystal surface.

To obtain the dispersion $E(k)$ in the forbidden gaps, let us insert (2.41) in Eq. (2.38). The energy $E(k)$ must be real. That leads to the following conditions on q and κ:

$$q = \pm\frac{\pi n}{a} \tag{2.44}$$

$$0 \leq \kappa \leq \frac{m_e a |V_n|}{\pi \hbar^2 n} \tag{2.45}$$

where $n = 1, 2, 3, \ldots$. The whole dispersion of $E(k)$ across the first allowed and forbidden zones is shown in Fig. 2.5.

2.1.2.3 Tamm and Shockley States

Let us consider the forbidden gap between the first and the second allowed energy bands in some more detail. In this case one has $n = \pm 1$ and quasi-degeneracy of the energies at the first Brillouin zone boundaries, $k = \pm\pi/a$.

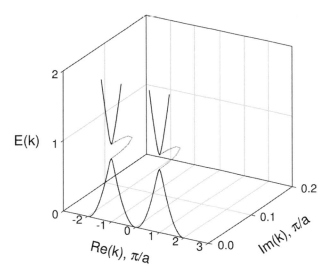

Fig. 2.5 The dispersion of electronic energy in a 1D crystal within the first and second Brillouin zones. The energy intervals corresponding to the bulk states (solid lines) form the allowed energy bands. They are separated by forbidden energy gaps where the bulk states do not exist, but the surface electronic states (dotted lines) can occur. Energy is calculated in units of $\pi^2\hbar^2/2m_e a^2$ for the case where $|V_1| = 0.1$.

Substituting Eq. (2.38) in Eq. (2.37) we obtain for $n = +1$ the wave function at the top of the forbidden gap

$$\psi^+(z) = A\left[e^{-i(\pi/a)z} + (V_1/|V_1|)e^{i(\pi/a)z}\right] \tag{2.46}$$

and at the bottom

$$\psi^-(z) = A\left[e^{-i(\pi/a)z} - (V_1/|V_1|)e^{i(\pi/a)z}\right] \tag{2.47}$$

For $n = -1$ the exponents change signs. If $V_1 > 0(< 0)$ then ψ^+ is an even (odd) function of z, i.e., it exhibits s-like (p-like) character. Analogously, ψ^- is an odd (even) function, i.e., it exhibits p-like (s-like) character. A forbidden energy gap in which the wave function has s-like character at its bottom and p-like character at its top is known as a *direct gap*. Conversely, if the wave function is of p-type at the bottom and of s-type at the top, a forbidden energy gap is called an *inverted gap* to stress the inverted order in comparison with the s- and p-levels of isolated atoms. Surface states found in a direct FEG are usually called *Tamm states* (Tamm 1932), whereas those in an inverted FEG are referred to as *Shockley states* (Shockley 1939). The latter case implies that when one considers a plot of electronic energy spectrum versus lattice constant, a, the AEBs originating from s and p atomic orbitals ($a \to \infty$) cross each other when a is varying to a finite value. After the crossing point two energy levels corresponding to surface electronic states are separated, one from each of the bands.

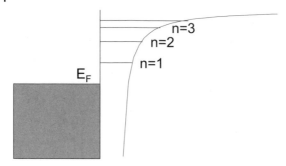

Fig. 2.6 Electron states in the image potential near a metal surface. E_F is the Fermi energy of the metal, the quantum number n numerates the hydrogen-like energy levels.

One can also consider this classification from the other point of view. An inverted gap implies that the formation of a crystal from single atoms (i.e., $a \to 0$ in the plot discussed above) leads to a strong perturbation of electronic states of a free atom. This case corresponds to the surface states considered by Shockley. On the contrary, Tamm first found surface states in the other limiting case of tightly bound electrons. However, there is no physical distinction between the different terms; only the mathematical approach is different. An example of Tamm states is found at semiconductor surfaces. The sp^3 hybrid orbital is responsible for chemical bonding in Si or Ge. When these bonds are broken on crystal cleavage, the remaining lobes stick out of the surface, giving the dangling bonds.

2.1.2.4 Image States

We have assumed above that the potential of an electron outside the crystal is constant (V_s). A more realistic approach has to take into account the dependence of the potential on the electron–surface distance. In the case of a metal surface, the Coulomb potential of the electron mirror image relative to the surface plane dictates the interaction,

$$V(z) = -\frac{e^2}{2z} \tag{2.48}$$

where e is the electron charge. The potential (2.48) generates a Rydberg series of surface states, which are similar to the states of the hydrogen-like atoms (Fig. 2.6). They are not related to the crystal termination. Being located above the Fermi level, the image states are unoccupied in equilibrium. However, they can be populated by light of an appropriate frequency. Then electrons in such states can propagate only along the metal surface, i.e., they are essentially electronic surface states.

2.1.2.5 **Surface States of a 3D Crystal**

The results obtained so far can be generalized to the 2D surface of a 3D crystal. Due to the 2D periodicity along the surface, the Bloch-type wave functions (2.19) have the form

$$\psi_{\mathbf{k}_{\parallel}}(\mathbf{r}) = u_{\mathbf{k}_{\parallel}}(\mathbf{r})e^{i\mathbf{k}_{\parallel}\cdot\mathbf{r}_{\parallel}} \tag{2.49}$$

where $\mathbf{k}_{\parallel}$ and $\mathbf{r}_{\parallel}$ are the components of the wave vector and the radius vector, respectively, parallel to the surface. The function $u_{\mathbf{k}_{\parallel}}$ is invariant with respect to any translation parallel to the surface, i.e.,

$$u_{\mathbf{k}_{\parallel}}(\mathbf{r} + \mathbf{a_n}) = u_{\mathbf{k}_{\parallel}}(\mathbf{r}) \tag{2.50}$$

where $\mathbf{a_n}$ is a vector belonging to the 2D Bravais lattice.

The energies of the electronic surface state, $E_s(\mathbf{k}_{\parallel})$, are now represented by surfaces in 3D space. To image the electronic states at the surface, one has also to consider the bulk states, $E_b(\mathbf{k}_{\parallel}, k_{\perp})$ with $k_{\perp}$ the wave vector component perpendicular to the surface, corresponding to electrons propagating to the surface and away from it. This can be done by means of projecting the surfaces $E_b(\mathbf{k}_{\parallel}, k_{\perp})$ onto the 2D reciprocal $\mathbf{k}_{\parallel}$-space. As a simplification one usually considers a contour connecting points of high symmetry in the $\mathbf{k}_{\parallel}$-space and plots both $E_s(\mathbf{k}_{\parallel})$ and $E_b(\mathbf{k}_{\parallel}, k_{\perp})$ along this contour. We thus come to the presentation of electronic states as continuous allowed bands for the bulk states (arising from different $k_{\perp}$) and as discrete lines disposed between them for the surface states (Fig. 2.7). In a 3D crystal a surface state can, in principle, be degenerate with the bulk states for some values of $\mathbf{k}_{\parallel}$. In such a case, an electron being in the surface state has a certain probability to transfer into the bulk state and to penetrate deep into the crystal. These states are referred to as *surface resonances*.

2.1.2.6 **Extrinsic Surface States**

The surface states considered so far are related to the clean and well-ordered surface of a crystal and they are called therefore *intrinsic surface states*. Defects or impurities located at a surface can form electronic states whose wave functions are localized in their vicinity and thus at the surface too. They are called *extrinsic surface states*. Such states do not possess a 2D translational symmetry[2] and an electron occupying such a state cannot propagate along the surface.

2.1.3
Surface Plasmons

Collective excitations of conduction electrons localized near the conductor surface are called *surface plasmons*. In the nonretarded limit ($c \to \infty$) or, equiv-

---

2) An exception is the case of a 2D lattice of adsorbed atoms or molecules.

(a)   (b)

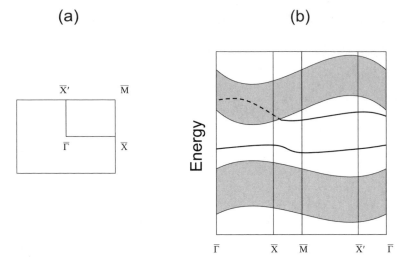

Fig. 2.7 (a) The first 2D Brillouin zone of a (110) surface of a cubic crystal. (b) Schematic representation of the electronic energy $E(\mathbf{k}_\parallel)$ along the symmetry lines. The shaded areas show the projections of the bulk allowed energy bands. The solid and dashed lines represent the surface electronic states and surface resonance, respectively.

alently, in the long-wavelength limit ($\omega \to 0$) one can omit, in Maxwell's equations, the term $(1/c)(\partial \mathbf{H}/\partial t)$ and thus reduce the problem to the electrostatic case. Then from the equations

$$\nabla \times \mathbf{E} = 0 \tag{2.51}$$

and

$$\nabla \cdot \mathbf{E} = 0 \tag{2.52}$$

we conclude that the electric field of a surface plasmon can be found as

$$\mathbf{E} = -\nabla \varphi \tag{2.53}$$

where the potential φ satisfies the Laplace equation,

$$\Delta \varphi = 0 \tag{2.54}$$

It can be described by a wave-like form on both sides of the surface:

$$\varphi_1(\mathbf{r}_\parallel, z) = \varphi_{10} e^{-\kappa_1 z} e^{i\mathbf{k}_\parallel \cdot \mathbf{r}_\parallel} e^{-i\omega t}, \quad z > 0 \tag{2.55}$$

$$\varphi_2(\mathbf{r}_\parallel, z) = \varphi_{20} e^{\kappa_2 z} e^{i\mathbf{k}_\parallel \cdot \mathbf{r}_\parallel} e^{-i\omega t}, \quad z < 0 \tag{2.56}$$

where $\kappa_1, \kappa_2 > 0$. The substitution of (2.55) and (2.56) into Eq. (2.54) gives the relations

$$\kappa_1 = \kappa_2 = k_\parallel \tag{2.57}$$

From here one finds the electric field amplitudes parallel to the surface:

$$\mathbf{E}_{10,\parallel}(\mathbf{r}_{\parallel}, z = 0^+) = -i\mathbf{k}_{\parallel}\varphi_{10}e^{i\mathbf{k}_{\parallel}\cdot\mathbf{r}_{\parallel}} \tag{2.58}$$

$$\mathbf{E}_{20,\parallel}(\mathbf{r}_{\parallel}, z = 0^-) = -i\mathbf{k}_{\parallel}\varphi_{20}e^{i\mathbf{k}_{\parallel}\cdot\mathbf{r}_{\parallel}} \tag{2.59}$$

and perpendicular to the surface

$$E_{10,\perp}(\mathbf{r}_{\parallel}, z = 0^+) = -k_{\parallel}\varphi_{10}e^{i\mathbf{k}_{\parallel}\cdot\mathbf{r}_{\parallel}} \tag{2.60}$$

$$E_{20,\perp}(\mathbf{r}_{\parallel}, z = 0^-) = k_{\parallel}\varphi_{20}e^{i\mathbf{k}_{\parallel}\cdot\mathbf{r}_{\parallel}} \tag{2.61}$$

Finally, the continuity of the **E**-vector tangential component and the **D**-vector normal component across the surface leads to the equations

$$\varphi_{10} = \varphi_{20} \tag{2.62}$$

and

$$\epsilon(\omega) = -1 \tag{2.63}$$

with $\epsilon(\omega)$ the dielectric function of the crystal. Equation (2.63) determines the frequency of surface plasmon, ω_s, in the long-wavelength limit. In the Drude model[3]

$$\epsilon(\omega) = 1 - \frac{\omega_p^2}{\omega^2} \tag{2.64}$$

where

$$\omega_p = \sqrt{\frac{4\pi n e^2}{m^*}} \tag{2.65}$$

is the plasma frequency with n the concentration of conduction electrons, e and m^* the electron charge and effective mass, respectively. Then one obtains

$$\omega_s = \frac{\omega_p}{\sqrt{2}} \tag{2.66}$$

The jump of the electric field component normal to the surface gives the surface density of a charge located at $z = 0$ which corresponds to a surface plasmon:[4]

$$\sigma = \frac{1}{4\pi}[E_{10,\perp}(\mathbf{r}_{\parallel}, z = 0^+) - E_{20,\perp}(\mathbf{r}_{\parallel}, z = 0^-)] = -\frac{k_{\parallel}\varphi_{10}}{2\pi}e^{i\mathbf{k}_{\parallel}\cdot\mathbf{r}_{\parallel}} \tag{2.67}$$

2.1.4
Surface Phonons

So far we have discussed surface properties for the case where the atoms which constitute a crystal occupy their equilibrium positions. Taking into account of the kinetic energy of nuclei leads to the equation for their vibrational

---

3) For the sake of simplicity, we have neglected here the scattering of electrons.

4) Only the real part of this expression has a physical sense.

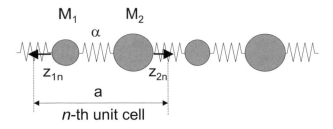

Fig. 2.8 Model of a diatomic one-dimensional crystal.

motion in which the electronic energy plays the role of a potential. Since the electronic energy at the surface differs from that in the crystal bulk, the frequencies of surface vibrational excitations (*surface phonons*) differ from the bulk phonon spectrum.

The essential features of surface vibrational dynamics can be analyzed, as it has been done for surface electronic states, by considering a 1D semi-infinite chain of atoms. We assume for simplicity that the chain consists of two kinds of atoms with masses M_1 and M_2 (see Fig. 2.8) and only neighbouring atoms interact with each other through the elastic forces

$$F_n = -\alpha \Delta z_n \tag{2.68}$$

Here, α is the force constant and Δz_n is the difference between the displacements z_{1n} and z_{2n} of two adjacent atoms from their equilibrium positions, where the subscript n specifies the number of the chain unit cell. Then Newton's second law reads

$$M_1 \ddot{z}_{1n} = -\alpha(z_{1n} - z_{2n}) - \alpha(z_{1n} - z_{2,n+1})$$
$$= -\alpha(2z_{1n} - z_{2n} - z_{2,n+1}) \tag{2.69}$$

$$M_2 \ddot{z}_{2n} = -\alpha(2z_{2n} - z_{1n} - z_{1,n-1}) \tag{2.70}$$

We shall seek the solutions of Eqs (2.69) and (2.70) in a form which corresponds to collective vibrations of atoms:

$$z_{jn} = A_j e^{ikna} e^{-i\omega t} \quad j = 1, 2 \tag{2.71}$$

where k and ω are the wave number and the frequency of the vibration, a is the distance between the equilibrium positions of atoms of the same kind and A_j are the constants to be determined from Newton's equations. Substitution of (2.71) into Eqs (2.69) and (2.70) leads to a system of two linear homogeneous equations with respect to A_1 and A_2:

$$\begin{pmatrix} M_1\omega^2 - 2\alpha, & \alpha(1 + e^{ika}) \\ \alpha(1 + e^{-ika}), & M_2\omega^2 - 2\alpha \end{pmatrix} \begin{pmatrix} A_1 \\ A_2 \end{pmatrix} = 0 \tag{2.72}$$

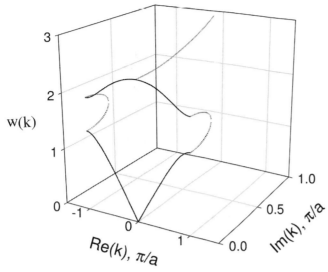

Fig. 2.9 Dispersion branches of "bulk" (solid lines) and "surface" (dotted lines) phonons of a 1D semi-infinite crystal. The frequency is calculated for the case where $M_2 = 2M_1$ in units of $(\alpha / M_1)^{1/2}$

This system has nontrivial solutions only if its determinant is equal to zero. Thus we obtain for the allowed frequencies of collective vibrations

$$\omega_{\pm}^2(k) = \frac{\alpha}{M_1 M_2} \left(M_1 + M_2 \pm \sqrt{M_1^2 + M_2^2 + 2 M_1 M_2 \cos ka} \right) \qquad (2.73)$$

The solutions $\omega_-(k)$ and $\omega_+(k)$ correspond to the acoustic and optical phonon branches, respectively, in an infinite 1D chain of atoms (Fig. 2.9). To ensure the finiteness of the atomic displacements given by Eq. (2.71) one has to require that the wave number k is real. However, this is not necessary if we consider an atomic chain which is terminated at one end. In that case k has an imaginary part $i\kappa$ (Eq. (2.41)) and, accordingly,

$$\cos ka = \cos qa \cosh \kappa a - i \sin qa \sinh \kappa a \qquad (2.74)$$

Thus $\omega_{\pm}(k)$ are real if the equation

$$\sin qa \sinh \kappa a = 0 \qquad (2.75)$$

is satisfied. For $\kappa \neq 0$ we obtain from here $q = \pi m/a$ with $m = 0, \pm 1, \pm 2, \ldots$. These solutions with $m \neq 0$ correspond to the Brillouin zone boundaries. For the first Brillouin zone ($m = 0, \pm 1$)

$$\cos ka = \cos \pi m \cosh \kappa a = (-1)^m \cosh \kappa a, \quad m = 0, \pm 1 \qquad (2.76)$$

24 | 2 Surfaces and Interfaces

and therefore

$$\omega_+^2(i\kappa) = \frac{\alpha}{M_1 M_2}\left(M_1 + M_2 + \sqrt{M_1^2 + M_2^2 + 2M_1 M_2 \cosh \kappa a}\right) \quad (2.77)$$

and

$$\omega_\pm^2\left(\frac{\pi}{a} + i\kappa\right) = \frac{\alpha}{M_1 M_2}\left(M_1 + M_2 \pm \sqrt{M_1^2 + M_2^2 - 2M_1 M_2 \cosh \kappa a}\right) \quad (2.78)$$

Note that only the plus sign in front of the square-root in Eq. (2.77) is taken to ensure the realness of the frequency and that the solution for $k = -(\pi/a) + i\kappa$ coincides with (2.78). The corresponding 1D surface phonon branches are shown in Fig. 2.9. The possible frequencies of surface vibrations fill a range which does not overlap with the acoustic and optical branches found for the bulk excitations.

Substitution of (2.41) into Eq. (2.71) gives

$$z_{jn} = (-1)^{mn} A_j e^{-\kappa n a} e^{-i\omega t} \quad j = 1,2; \quad m = 0, \pm 1, \ldots \quad (2.79)$$

i.e., the amplitudes of the surface vibrations decrease exponentially with increasing number of the unit cell, n (i.e., with distance from the surface).

The results obtained above for a 1D chain of atoms can be qualitatively extended to the case of a 3D crystal terminated by a 2D surface, much as it has been demonstrated before for electronic surface states. One can imagine a 3D solid as being constructed of an infinite number of linear chains parallel to each other. Due to the interatomic interactions the phases of vibrations in different chains are correlated to each other. A surface phonon then can be described by a wave vector $\mathbf{k}_\parallel$ parallel to the surface. The atomic displacements are now represented by the vectors

$$\mathbf{z}_{jn} = A_j \hat{\mathbf{e}}_j e^{-\kappa n_3 a_{3z}} e^{i\mathbf{k}_\parallel \cdot (n_1 \mathbf{a}_1 + n_2 \mathbf{a}_2)} e^{-i\omega t} \quad (2.80)$$

where $\hat{\mathbf{e}}_j$ is the unit vector along the displacement of the jth atom in the unit cell and it is implied that the 3D Bravais lattice is determined by the vectors

$$\mathbf{a_n} = n_1 \mathbf{a}_1 + n_2 \mathbf{a}_2 + n_3 \mathbf{a}_3 \quad (2.81)$$

with the primitive vectors $\mathbf{a}_1$ and $\mathbf{a}_2$ being parallel to the surface plane.

The surface phonon dispersion relation $\omega(\mathbf{k}_\parallel)$ is represented by a surface in 3D space. Usually one displays it by plotting the function $\omega(\mathbf{k}_\parallel)$ along certain symmetry lines of the 2D Brillouin zone together with the dispersion relation $\omega(\mathbf{k}_\parallel, k_\perp)$ for bulk phonons. Then the surface phonon dispersion is given by a line whereas the bulk phonon frequencies fill an area originating from different values of $k_\perp$ (Fig. 2.10).

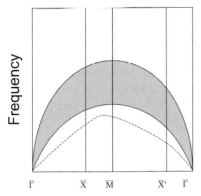

Fig. 2.10 Schematic representation of the bulk (shaded area) and surface (dashed line) phonon dispersions for a (110) surface of a cubic crystal. The symmetry lines for the first 2D Brillouin zone are shown in Fig. 2.7a.

Among surface phonon branches one can distinguish that of the surface optical phonon, $\omega_s(\mathbf{k}_\parallel)$, which corresponds to surface vibrations of ionic crystals accompanied by an oscillating dipole moment. Its frequency in the long-wavelength limit ($\mathbf{k}_\parallel \to 0$) can be found, much in the same way as for the case of surface plasmons (see Section 2.1.3). The corresponding macroscopic electric field is determined by an electrostatic potential of the forms (2.55) and (2.56). As a result, ω_s satisfies the equation $\epsilon(\omega_s) = -1$, where $\epsilon(\omega)$ is the dielectric function of the crystal. Taking the latter quantity in the form

$$\epsilon(\omega) = \epsilon_\infty + (\epsilon_0 - \epsilon_\infty)\frac{\omega_{TO}^2}{\omega_{TO}^2 - \omega^2} \tag{2.82}$$

with ϵ_0 and ϵ_∞ the static and high-frequency dielectric constants, respectively, and ω_{TO} the bulk optical phonon frequency, one finds

$$\omega_s = \omega_{TO}\sqrt{\frac{\epsilon_0 + 1}{\epsilon_\infty + 1}} \tag{2.83}$$

2.1.5
Surface Roughness

In the preceding sections we have considered a well-defined surface structure typical for crystal faces prepared under ultra-high-vacuum (UHV) conditions. In practice, one more frequently deals with surfaces which are rough. These require a different description.

One does not find two rough surfaces which are identical and even those which were obtained under the same processing parameters may differ from each other in an uncontrollable way. Such surfaces can be treated in a similar way to statistical random processes. A rough surface is described in terms of its deviation from a smooth "reference surface". The surface relief is represented by the heights of its points as a function of their positions along the reference plane, $\zeta(\mathbf{r}_\parallel)$. The probability for a surface point to be at a height between ζ and $\zeta + d\zeta$ is given by $p(\zeta)d\zeta$, where $p(\zeta)$ is the height distribution function. Without any loss of generality one can choose the reference surface in a way that fulfils the condition

$$\langle \zeta \rangle = \int_{-\infty}^{\infty} \zeta p(\zeta) d\zeta = 0 \tag{2.84}$$

where the angular brackets denote averaging over the surface area. Then the root-mean-square height of the surface is equal to the standard deviation and is given by

$$\sigma = \sqrt{\langle \zeta^2 \rangle} \tag{2.85}$$

In many cases the height distribution can be well described by the Gaussian function

$$p(\zeta) = \frac{1}{\sigma\sqrt{2\pi}} \exp\left(-\frac{\zeta^2}{2\sigma^2}\right). \tag{2.86}$$

An additional feature of a rough surface related to the length scale over which the height changes, is the correlation function

$$C(\mathbf{r}_\parallel' - \mathbf{r}_\parallel) = \langle \zeta(\mathbf{r}_\parallel)\zeta(\mathbf{r}_\parallel') \rangle. \tag{2.87}$$

It has the property that $C(\mathbf{0}) = \sigma^2$. When $\mathbf{r}_\parallel' - \mathbf{r}_\parallel$ increases, $C(\mathbf{r}_\parallel' - \mathbf{r}_\parallel)$ decays to zero for a random surface with the rate depending on the distance over which points become uncorrelated. Often it is assumed that the correlation function is also Gaussian, i.e.,

$$C(|\mathbf{r}_\parallel' - \mathbf{r}_\parallel|) = \sigma^2 \exp\left(-\frac{|\mathbf{r}_\parallel' - \mathbf{r}_\parallel|^2}{\lambda_0^2}\right) \tag{2.88}$$

where we imply that the surface is isotropic, so only the modulus of $\mathbf{r}_\parallel' - \mathbf{r}_\parallel$ enters the equation. The distance λ_0 at which the correlation function falls by $1/e$ is called the *correlation length*. Figure 2.11 shows surface profiles of the same σ but with different correlation lengths. Obviously, an appropriate description of surface roughness has to present values of both σ and λ_0.

One can introduce also the *characteristic function* of a rough surface which is the Fourier transform of the height distribution function,

$$\chi(s) = \int_{-\infty}^{\infty} p(\zeta)e^{-is\zeta}d\zeta \tag{2.89}$$

Although $\chi(s)$ does not contain any additional information in comparison with the function $p(\zeta)$, it is of use in the theory of light scattering by rough (a)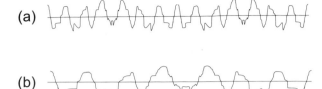

(b)

(c)

Fig. 2.11 Surface profiles of the same RMS height but with the correlation lengths increasing from (a) to (c).

surfaces. Analogously, the Fourier transform of the correlation function gives a new characteristic of a rough surface called the *power spectrum*:

$$P(\mathbf{k}_\|) = \int C(\mathbf{R})e^{-i\mathbf{k}_\| \cdot \mathbf{R}} d\mathbf{R} \tag{2.90}$$

The power spectrum can be expressed in terms of the surface profile by substituting Eq. (2.87) into (2.90):

$$P(\mathbf{k}_\|) = \lim_{A \to \infty} \frac{1}{A} \left| \hat{\zeta}(\mathbf{k}_\|) \right|^2 \tag{2.91}$$

where

$$\hat{\zeta}(\mathbf{k}_\|) = \int_A \zeta(\mathbf{r}_\|)e^{-i\mathbf{k}_\| \cdot \mathbf{r}_\|} d\mathbf{r}_\| \tag{2.92}$$

with A the area of the mean (reference) surface and $\hat{\zeta}(-\mathbf{k}_\|) = \hat{\zeta}^*(\mathbf{k}_\|)$ because of the realness of ζ. In the limit $A \to \infty$ Eq. (2.92) provides the Fourier transform of the surface profile. Note that the power spectrum describes both the spread of heights about the reference plane and the height variation along the surface.

2.2
Adsorption on Solid Surfaces

So far we have considered the structure and properties of solid surfaces which are clean, i.e., they do not contain any foreign atoms or molecules. Such surfaces can be prepared under UHV conditions. However, in many practical cases, surfaces contain atoms and molecules from the surrounding gas phase which form more or less strong bonds with surface atoms. This phenomenon is called *adsorption*. Depending on the nature of the forces between the gas atoms or molecules (*adsorbate*) and the surface (*adsorbent*) one recognizes physical adsorption (or *physisorption*) and chemical adsorption (or *chemisorption*).

2.2.1
Physisorption

If the coupling between a foreign atom or a molecule and a surface does not involve the formation of a chemical bond, the adsorption is referred to as physisorption. The underlying mechanism is known in molecular physics as a *van der Waals interaction* and is due to the interaction between instantaneous dipoles which are generated in interacting molecules because of the quantum-mechanical fluctuations of their electron clouds. The van der Waals potential is attractive and decreases with distance r between the molecules as

$$V(r) = -\frac{C_6}{r^6} \tag{2.93}$$

where C_6 is the first-order (dipole) interaction constant.

Let us have a look at the image charges induced by an atomic dipole in the polarizable substrate. According to classical electrodynamics (Stratton 1941), the electrostatic field of a charge q located at a distance z from a flat surface of a medium with real dielectric function ϵ can be described by introducing a charge

$$q' = -\frac{\epsilon - 1}{\epsilon + 1}q \tag{2.94}$$

located at the position of the mirror image. As a consequence, an electric dipole $\vec{\mu} = (\vec{\mu}_\parallel, \vec{\mu}_\perp)$ "creates" an image dipole with the components

$$\vec{\mu}'_\parallel = -\frac{\epsilon - 1}{\epsilon + 1}\vec{\mu}_\parallel \tag{2.95}$$

and

$$\vec{\mu}'_\perp = \frac{\epsilon - 1}{\epsilon + 1}\vec{\mu}_\perp \tag{2.96}$$

Note that the components of the dipole moments parallel to the surface ($\parallel$) are antiparallel to each other, whereas those perpendicular to the surface ($\perp$) are of the same direction (see Fig. 2.12). The interaction energy of the "real" dipole with its image is then determined as (Landau and Lifshitz 1980)

$$V(z) = \frac{(\vec{\mu} \cdot \vec{\mu}') - 3(\vec{\mu} \cdot \hat{\mathbf{n}})(\vec{\mu}' \cdot \hat{\mathbf{n}})}{(2z)^3} \tag{2.97}$$

where $\hat{\mathbf{n}}$ is the unit vector along the straight line between the dipoles. Substitution of (2.95) and (2.96) into Eq. (2.97) gives the result

$$V(z) = -\frac{\epsilon - 1}{\epsilon + 1}\frac{\mu_\parallel^2 + 2\mu_\perp^2}{8z^3} \tag{2.98}$$

In contrast to the van der Waals potential between two molecules, the atom–surface potential decreases as $1/z^3$ with distance from the surface.

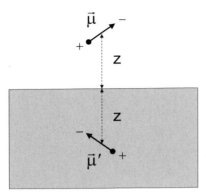

Fig. 2.12 The image dipole $\vec{\mu}'$ induced in a polarizable substrate by the "real" dipole $\vec{\mu}$.

Equation (2.98) describes the frequency shift of a classical harmonic oscillator, δE_a^{cl}, located above a surface at distances much less than the wavelength of its oscillations (Chance et al. 1975b). A rigorous quantum-electrodynamical treatment of the problem (Wylie and Sipe 1984, 1985) predicts, besides this classical term, a correction term arising from the van der Waals interaction between atom and surface. The total shift of the electronic energy level a can be written as

$$\delta E_a = \delta E_a^{vdW} + \delta E_a^{cl} \qquad (2.99)$$

Here, δE_a^{vdW} is the van der Waals shift, whereas the term δE_a^{cl} can be represented as the sum of the frequency shifts corresponding to classical harmonic oscillators associated with the dipole transitions to the lower atomic levels. In the case when a is the ground atomic state, $\delta E_a^{cl} = 0$. Therefore, the change in the energy of the ground state is a pure (non-classical) van der Waals shift.

The different contributions to the level shift can be found as

$$\delta E_a^{vdW} = -\frac{\hbar}{2\pi} \sum_j \int_0^\infty d\xi \, G_{jj}(i\xi) \alpha_{jj}^a(i\xi) \qquad (2.100)$$

and

$$\delta E_a^{cl} = -\sum_{nj} \mu_j^{an} \mu_j^{na} \operatorname{Re} G_{jj}(\omega_{na}) \Theta(\omega_{an}) \qquad (2.101)$$

with ω_{mn} and μ_j^{mn} the frequency and the dipole matrix element of the atomic transition $|m\rangle \rightarrow |n\rangle$ at infinite distance from the surface, $\Theta(\omega)$ the unit step function, and $j = x, y, z$. Here we have introduced the dynamic polarizability tensor of the ath atomic state

$$\alpha_{ij}^a(\omega) = \frac{2}{\hbar} \sum_n \frac{\omega_{na} \mu_i^{an} \mu_j^{na}}{\omega_{na}^2 - \omega^2} \qquad (2.102)$$

The functions $G_{ij}(\omega)$ describe the electric field reflected from the surface at the position of the dipole in terms of the dipole moment as

$$E_i^R(\omega) = \sum_j G_{ij}(\omega)\mu_j \tag{2.103}$$

In the limit $z \ll \lambda$ with λ the wavelength associated with the largest atomic transition frequency, i.e., neglecting retardation, the $G_{ij}(\omega)$s are found as

$$G_{xx}(\omega) = G_{yy}(\omega) = \frac{1}{8z^3}\frac{\epsilon(\omega)-1}{\epsilon(\omega)+1} \tag{2.104}$$

$$G_{zz}(\omega) = 2G_{xx}(\omega) \tag{2.105}$$

$$G_{ij}(\omega) = 0, \quad i \neq j \tag{2.106}$$

In the case of an atom above a metal surface described by the Drude dielectric function, Eq. (2.64), the classical term has the form

$$\delta E_a^{cl} = -\frac{1}{8z^3}\sum_n \frac{\Theta(\omega_{an})\omega_s^2[(\mu_\parallel^{na})^2 + 2(\mu_\perp^{na})^2]}{\omega_s^2 - \omega_{na}^2} \tag{2.107}$$

with ω_s the surface plasmon frequency, Eq. (2.66). If one of the atomic transition frequencies is close to ω_s, the term δE_a^{cl} shows a dispersion-like behavior and can have either positive or negative values depending on the sign of the frequency mismatch $\omega_{an} - \omega_s$. The result (2.107) has been obtained by neglecting the imaginary part of $\epsilon(\omega)$. If one takes it into account the divergence at the exact resonance $\omega_{an} = \omega_s$ does not occur.

The discussion above is valid at distances from the surface which are much larger than the Bohr radius, where the atom can be modelled as a point dipole. At shorter distances one has to use more adequate approximations which take into account the electronic charge distributions of both atom and surface and involve large-scale computer calculations. Qualitatively, in addition to the attractive forces discussed above, if an atom approaches the surface, its electron shell undergoes repulsion due to both Coulomb and exchange interactions with the substrate electrons. At some intermediate distance the atom–surface potential (*adsorption potential*) thus has a minimum. The interaction force varies along the surface following its atomic structure and hence the adsorption potential depends not only on the z-coordinate but also on the lateral position of an atom, $\mathbf{r}_\parallel$. Local minima of such a potential surface $V(\mathbf{r}_\parallel, z)$ correspond to equilibrium positions (*adsorption sites*) of an atom and their energy depths are given by the *binding* or *adsorption energy*. Physisorption is characterized by low binding energies, usually less than 0.3 eV, and by large atom–surface distances in the range 3–10 Å. Figure 2.13 shows a typical physisorption potential for $\mathbf{r}_\parallel$ fixed at an adsorption site.

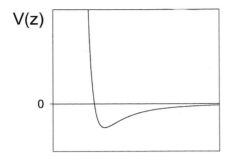

Fig. 2.13 Schematic cross-section of the physisorption potential $V(\mathbf{r}_\parallel, z)$ by the plane $\mathbf{r}_\parallel = const$ corresponding to an adsorption site. The minimum of this curve determines the equilibrium position and the binding energy of an adsorbed atom.

2.2.2
Chemisorption

A rigorous description of the formation of a chemical bond between an adsorbate and the surface would involve an infinite number of electrons in the substrate with a nearly continuous spectrum of energies. This is not a treatable computational problem. The next simpler approach would be to model the surface by a cluster of atoms. However, for a qualitative understanding of the nature of the chemical bonding at the surface one can also apply the concept of frontier orbitals.

The strongest interaction with an adsorbed molecule is assumed to occur via the molecular orbitals whose energy levels are nearest to the Fermi energy in the substrate. One of them, located below the Fermi energy, is called the *Highest Occupied Molecular Orbital (HOMO)*, the other one located above the Fermi level is called the *Lowest Unoccupied Molecular Orbital (LUMO)* (Fig. 2.14).

The energy of the molecule–surface bond is found from the variational principle by means of minimizing the energy functional

$$E(\psi) = \frac{\langle \psi | H | \psi \rangle}{\langle \psi | \psi \rangle} \tag{2.108}$$

where the trial wavefunction is taken in the form

$$\psi = N\psi_0(A, S) + \sum_{k<k_F} A_k \psi_k(A^-, S^+) + \sum_{k>k_F} B_k \psi_k(A^+, S^-) \tag{2.109}$$

and H is the Hamiltonian of the system "adsorbate (A) + substrate (S)". Here, ψ_0 is an unperturbed wavefunction of the system consisting of the neutral adsorbate and the neutral substrate, $\psi_k(A^-, S^+)$ and $\psi_k(A^+, S^-)$ are the unperturbed wavefunctions of the charge-transfer states corresponding to the transfer of an electron with the wave

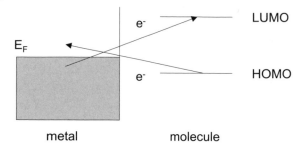

Fig. 2.14 Scheme illustrating the concept of the HOMO and LUMO orbitals at a metal surface.

number k from the substrate level below the Fermi energy to the LUMO, and from the HOMO to the level above the Fermi state in the substrate, respectively (Fig. 2.14). N, A_k and B_k are constants.

The result to second order in the interaction is given by

$$E - E_0 = \sum_{k < k_F} \frac{|U_{Lk}|^2}{\epsilon_k - \epsilon_{LUMO}} + \sum_{k > k_F} \frac{|U_{Hk}|^2}{\epsilon_{HOMO} - \epsilon_k} \tag{2.110}$$

with E_0 the energy of the nonbonding state, ϵ_{LUMO} and ϵ_{HOMO} the energies of the LUMO and HOMO states, U_{Lk} and U_{Hk} the interaction matrix elements between the substrate electron state ψ_k and the LUMO and HOMO states, respectively.

Introducing the spectral density functions

$$w_{LUMO}(\epsilon) = \sum_k |U_{Lk}|^2 \delta(\epsilon - \epsilon_k) \tag{2.111}$$

and

$$w_{HOMO}(\epsilon) = \sum_k |U_{Hk}|^2 \delta(\epsilon - \epsilon_k) \tag{2.112}$$

one can rewrite Eq. (2.110) in the form:

$$E - E_0 = 2 \int_{-\infty}^{\epsilon_F} d\epsilon \frac{w_{LUMO}(\epsilon)}{\epsilon - \epsilon_{LUMO}} + 2 \int_{\epsilon_F}^{0} d\epsilon \frac{w_{HOMO}(\epsilon)}{\epsilon_{HOMO} - \epsilon} \tag{2.113}$$

where the factor 2 takes into account the spin of the electrons. Both integrals on the right-hand side of Eq. (2.113) are negative and thus both contribute to the stabilization of the chemisorption state.

Chemisorption is characterized by binding energies of a few eV, i.e., much larger than in physisorption. The corresponding equilibrium distances are about 1–3 Å, i.e., shorter than for physically adsorbed species. If one draws on the same plot, the adsorption potentials for a physically adsorbed molecule and for chemically adsorbed atoms of which it consists, then the two potential curves cross each other at some intermediate distance from the surface (see Fig. 2.15). This means that a molecule approaching the surface has a certain probability to transfer to the other electronic term through a Landau–Zener

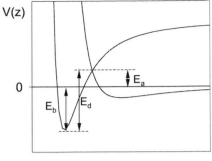

Fig. 2.15 Schematic representation of activated dissociative adsorption. A molecule approaching the surface is physisorbed. If its kinetic energy exceeds the potential barrier E_a, it can transfer to the chemisorption potential, which corresponds to adsorption of separated atoms.

transition (Landau and Lifshitz 1988). Such a transition is accompanied by dissociation of the molecule and formation of chemical bonds between the separated atoms and the surface. This process is called *dissociative adsorption*. In some cases the crossing point of the two potentials can be at a height E_a above the zero-energy level so that a potential barrier is formed (Fig. 2.15). Then the dissociative adsorption is *activated*, i.e., it becomes possible only for molecules having kinetic energies which exceed E_a (the *activation energy*). The reverse process, i.e., when the adsorbed atoms combine with each other and are desorbed as molecules is called *associative desorption*. The corresponding activation energy, E_d, is related to E_a by

$$E_d = E_b + E_a \tag{2.114}$$

with E_b the binding energy in the chemisorption potential.

2.2.3
Vibrational States of Adsorbates

2.2.3.1 Symmetry of Adsorption Sites
A free molecule consisting of N atoms has $3N$ degrees of freedom, 3 of which are translational, 3 (2 for a linear molecule) are rotational and the remaining $3N - 6$ ($3N - 5$) are vibrational. For a molecule adsorbed on a surface, translational and rotational motions are frustrated and all $3N$ normal coordinates correspond to vibrational modes. The symmetry of the corresponding vibrational Hamiltonian is dictated by the local symmetry of the adsorption site. The vibrational states of an adsorbate can be classified by irreducible representations of the point-group symmetry. To illustrate this principle, let us

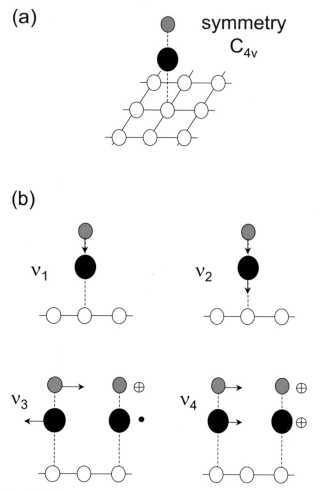

Fig. 2.16 Adsorption of a diatomic heteronuclear molecule at the (001) surface of a cubic crystal. (a) The on-top site characterized by the symmetry group C_{4v}. (b) The corresponding vibrational modes (side view). The degenerate modes are shown by pairs. The crossed circle means an arrow directed towards the plane of the picture, whereas a small dark circle means an arrow towards the viewer.

consider adsorption of a diatomic heteronuclear molecule on the (001) face of a simple cubic crystal (Richardson and Bradshaw 1979).

First, we assume that a molecule is adsorbed at the on-top site with its axis being perpendicular to the surface (Fig. 2.16). The point-group symmetry of this adsorbate–substrate complex is C_{4v} and the two stretching modes (ν_1 and ν_2) belong to the totally symmetric representation A_1, one of them (ν_2) originating from the translational motion along the surface normal. The remaining four modes are bending modes and correspond to two double-fold represen-

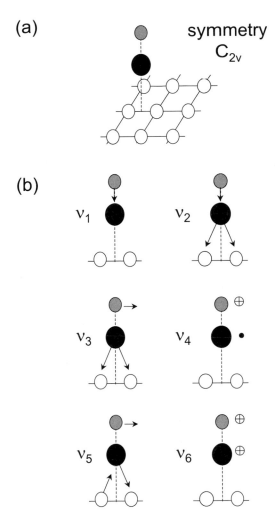

(a)

symmetry
C_{2v}

(b)

ν_1

ν_2

ν_3

ν_4

ν_5

ν_6

Fig. 2.17 Adsorption of a diatomic heteronu-clear molecule at the (001) surface of a cu-bic crystal. (a) The bridge site characterized by the symmetry group C_{2v}. (b) The corre-sponding vibrational modes (side view). The notations are the same as in Fig. 2.16. The vibrational modes in the second and third lines originate from the degenerate modes shown in Fig. 2.16 as a result of lowering of the symmetry.

tations E. Two of them are the frustrated translations along the crystal axes [100] and [010]. The other two vibrations originate from rotations of the mol-ecule around those axes.

This classification is changed when the same molecule is adsorbed at a bridge site between two surface atoms (Fig. 2.17). The new point-group sym-metry is C_{2v} and there are again two stretching vibrations corresponding to two A_1 representations. However, the degeneracy of the other modes is lifted, resulting in two vibrations in the plane of the bridge (ν_3 and ν_4) and two per-pendicular to it (ν_5 and ν_6).

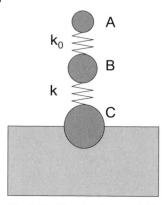

Fig. 2.18 The simplest model for the calculation of the vibrational frequency of an adsorbed diatomic molecule. The surface is replaced by a single atom C connected with the molecule AB by an elastic force with force constant k.

2.2.3.2 Vibrational Frequencies of Isolated Adsorbates

Except for a very few cases, the calculation of the vibrational frequencies of adsorbed molecules from first principles is not possible. Instead, one has to construct empirical models and one has to compare the result with experimental data.

In the simplest approach one can model the surface by a single atom which binds the molecule. Let us consider a diatomic molecule AB having an unperturbed eigenfrequency ω_0 which is attached to the surface atom C by an elastic force with the force constant k (Fig. 2.18). One obtains to first order in k a blue frequency shift due to adsorption ($\Delta\omega = \omega - \omega_0$)

$$\Delta\omega = \frac{k}{k_0} \frac{\mu^2}{2M_C^2} \omega_0 \qquad (2.115)$$

where k_0 is the force constant of the elastic force between the atoms A and B, μ is the reduced mass of the molecule AB given by

$$\mu = \frac{M_A M_B}{M_A + M_B} \qquad (2.116)$$

and $M_\alpha, \alpha = A, B, C$ are the masses of the corresponding atoms.

In a more sophisticated model, the substrate has one acoustic band $0 < \omega < \omega_D$[5] and the molecule has two or more unperturbed mode frequencies ω_{i0}. In many practical situations, light molecules are adsorbed on a crystal which consists of heavy atoms. Then the inner molecular frequencies ω_{i0} are all much higher than the band edge ω_D. In such a case, one can expand the result in a power series in the ratio of masses. For the linear quasimolecule

---

5) The frequency ω_D has the sense of the Debye frequency of a solid.

$AB - C$ with the stretching vibrations normal to the surface, one finds the blue frequency shifts

$$\Delta\omega_1 = \frac{M_A + M_B}{2M_C}\omega_{10} \tag{2.117}$$

for the $B - C$ bond, and

$$\Delta\omega_2 = \frac{M_A(M_A + M_B)}{2M_B M_C}\left(\frac{\omega_{10}}{\omega_{20}}\right)^4 \omega_{20} \tag{2.118}$$

for the $A - B$ bond.

Besides the mechanical renormalization considered above, which leads to an increase in the adsorbate vibrational frequency, there may be also a frequency shift due to the interaction of the molecular dipole with its image in the substrate. Such a shift has the same origin as that discussed for the electronic states in physisorption (Section 2.2.1). Applying that theory to the model of a one-dimensional harmonic oscillator, one finds that the polarizability α^a (Eq. (2.102)) is independent of the state $|a\rangle$, and thus the van der Waals shift δE_a^{vdW} (Eq. (2.100)) is the same for each state. As a result, the frequency shift is given by the difference between the classical correction terms δE_a^{cl} (Eq. (2.101)) corresponding to the ground $|0\rangle$ and first excited state $|1\rangle$ of a quantum oscillator (Wylie and Sipe 1985):

$$\Delta\omega = -\frac{1}{8\hbar z^3}[(\vec{\mu}_{\parallel}^{10})^2 + 2(\mu_{\perp}^{10})^2]\,\mathrm{Re}\left[\frac{\epsilon(\omega_0) - 1}{\epsilon(\omega_0) + 1}\right] \tag{2.119}$$

where $\epsilon(\omega_0)$ is the substrate dielectric function taken at the unperturbed oscillator frequency. This shift is negative if ω_0 is far from the substrate excitation frequencies, but may also be positive when $\epsilon(\omega_0)$ falls into the range between -1 and $+1$. The latter condition can be fulfilled, e.g., if ω_0 is close to the substrate phonon frequencies.

2.2.3.3 Coupled Vibrations of Overlayers

We have considered above the mechanisms which induce a frequency shift of an isolated molecule adsorbed on the surface. When the surface coverage increases, the oscillations of different molecular dipoles within the overlayer become correlated with each other, leading to a shift in the vibrational frequency of the molecular array with respect to the frequency of a single molecule.

One of the possible mechanisms for such a correlation is the dipole–dipole interaction between molecules. The wavelength λ of vibrations is usually of the order of a few microns. When the surface area of size $\sim \lambda^2$ contains a large number of adsorbed molecules, i.e.,

$$\lambda^2 N_s \gg 1 \tag{2.120}$$

with N_s the surface number density of molecules, all dipoles within this area that are resonantly excited by an external field, oscillate in phase with each other. Let us assume that all the adsorbed molecules are identical and that they form a 2D lattice described by vectors $\mathbf{b_m}$ of the form (2.4). Also, their dipole moments are assumed to be perpendicular to the surface. Then the local electric field at the molecule's site is the sum of the external field $E_{ex}(\omega) = E_0 \exp(i\mathbf{k} \cdot \mathbf{r} - i\omega t)$ and the dipole fields of all other molecules. The self-consistent equation for the field component normal to the surface is then given by

$$E_{\mathbf{m}}(\omega) = E_0 \exp(i\mathbf{k}_{\parallel} \cdot \mathbf{b_m} - i\omega t) - \sum_{n \neq m} \frac{\alpha_{zz}(\omega)E_{\mathbf{n}}(\omega)}{|\mathbf{b_m} - \mathbf{b_n}|^3} \qquad (2.121)$$

with $\alpha_{zz}(\omega)$ the dynamic polarizability component of a molecule in its ground state. Equation (2.121) can be solved using the substitution

$$E_{\mathbf{m}} = E \exp(i\mathbf{k}_{\parallel} \cdot \mathbf{b_m} - i\omega t) \qquad (2.122)$$

which gives

$$E = \frac{E_0}{1 + \alpha_{zz}(\omega)T(\mathbf{k}_{\parallel})} \qquad (2.123)$$

where

$$T(\mathbf{k}_{\parallel}) = \sum_{m \neq 0} \frac{\exp(i\mathbf{k}_{\parallel} \cdot \mathbf{b_m})}{|\mathbf{b_m}|^3} \qquad (2.124)$$

The poles of the solution (2.123) determined by the equation

$$1 + \alpha_{zz}(\omega)T(\mathbf{k}_{\parallel}) = 0 \qquad (2.125)$$

correspond to the collective normal modes of the adsorbed layer. In the long-wavelength limit, $\mathbf{k}_{\parallel} \to 0$, which corresponds to the condition (2.120), the summation over the 2D lattice in (2.124) converges to

$$T(\mathbf{0}) = CN_s^{3/2} \qquad (2.126)$$

with $C \approx 9$.

Assuming, for the molecular vibrations, the harmonic oscillator model, we can write the molecular dynamic polarizability as

$$\alpha(\omega) = \frac{\alpha_v}{1 - \omega^2/\omega_0^2} \qquad (2.127)$$

with $\alpha_v = \alpha(0)$ the static polarizability. Substituting this expression into Eq. (2.125), we find

$$\omega^2(\mathbf{k}_{\parallel}) = \omega_0^2[1 + \alpha_v T(\mathbf{k}_{\parallel})] \qquad (2.128)$$

Equation (2.128) is the dispersion relation for the collective vibrations of adsorbed molecules arranged in a 2D lattice, i.e., for 2D phonons of the overlayer.

The model considered above can be improved if one takes into account the electronic contribution, α_e, to the polarizability as well as the renormalization originating from the interaction of a molecule with its image in the substrate. Then instead of Eq. (2.128) one has the following dispersion law:

$$\omega^2(\mathbf{k}_\parallel) = \omega_0^2 \left[1 + \frac{\bar{\alpha}_v T(\mathbf{k}_\parallel)}{1 + \bar{\alpha}_e T(\mathbf{k}_\parallel)} \right] \tag{2.129}$$

where in the case of an overlayer on a perfect conductor surface

$$\bar{\alpha}_e = \frac{\alpha_e}{1 - (\alpha_e/4z^3)} \tag{2.130}$$

and

$$\bar{\alpha}_v = \frac{\alpha_v}{[1 - (\alpha_e/4z^3)]^2} \tag{2.131}$$

There may also be another mechanism of the frequency shift of the adsorbed layer due to short-range binding between molecules. The wavefunctions of closely located molecules overlap, leading to mutual influence on their vibrations. This effect can be considered in a similar way as for dipole–dipole interactions but in this case only adjacent molecules interact with each other. As a result, the overlayer phonons have two branches

$$\omega_\pm^2(\mathbf{k}_\parallel) = \omega_0^2 \pm |\Delta(\mathbf{k}_\parallel)| \tag{2.132}$$

with

$$\Delta(\mathbf{k}_\parallel) = \Delta \sum_{\mathbf{m} \neq 0}{}' \exp(i\mathbf{k}_\parallel \cdot \mathbf{b_m}) \tag{2.133}$$

where $\sum'$ implies the summation over the lattice points nearest to $\mathbf{m} = 0$ and the constant Δ characterizes the interaction between adjacent molecules.

2.2.4
Relaxation of Adsorbate Excitations

An adsorbed atom or a molecule being in its excited state is characterized by a finite lifetime which is determined by the reciprocal of the decay rate of this state. The finiteness of the lifetime leads to a broadening of the lines in the optical spectra of the adsorbate. Besides spontaneous emission which occurs also for free atoms and molecules, adsorbed species have other specific channels of relaxation, conditioned by their proximity to the surface. Any relaxation process must obey the conservation law of energy and therefore it takes place only if there is a substrate excitation which can accept the energy that the excited adsorbate releases. Therefore, possible decay mechanisms are determined by the energy spectrum of the substrate and thus generally are different for metals, semiconductors and dielectrics. They can be broadly classified as being mediated by photons, phonons, electron–hole pairs and conduction electrons.

Dephasing is another important broadening process for spectral lines of adsorbates. Elastic collisions of phonons and conduction electrons with adsorbed atoms or molecules disrupt the phases of their induced dipole moments and thus provide surface-specific pathways for phase relaxation. If an adsorbed particle can be considered as a two-level system, both the lifetime of its excited state, T_1^s, and the dephasing time, $T_2'^s$, contribute to the spectral linewidth $\gamma_\perp^s$ as[6]

$$\gamma_\perp^s \equiv \frac{2}{T_2^s} = \frac{1}{T_1^s} + \frac{2}{T_2'^s} \tag{2.134}$$

In the following we shall consider the different relaxation pathways in some detail.

Photons

The radiative lifetime ranges from 10^{-7} to 10^{-5} s for atomic transitions and from 10^{-3} to 0.1 s for vibrational transitions in molecules. Therefore, radiative decay is entirely negligible in comparison with the other pathways on metal and semiconductor surfaces. However, it can play a role in relaxation of electronically excited states of adsorbates on dielectrics.

Phonons

This mechanism concerns the relaxation of vibrationally excited molecules near surfaces. It depends strongly on the ratio between the vibrational frequency of a molecule, ω_0, and the Debye frequency of the substrate, ω_D, which determines the upper limit of the phonon spectrum. If $\omega_0 < \omega_D$, relaxation through creation of a single phonon is possible. Usually the corresponding decay rates are of the order of 10^{12}–10^{13} s^{-1}. For $(n-1)\omega_D < \omega_0 < n\omega_D$, the relaxation is accompanied by the generation of n phonons. The probability of an n-phonon process to occur rapidly decreases with order n. Typical values for the two-, three- and four-phonon decay rates are 10^{10}–10^{11}, 10^8–10^9 and 10^6–10^8 s^{-1}, respectively (Zhdanov and Zamaraev 1982).

Electron–hole pairs

In the case of adsorption on a metal surface, there may be an additional channel of relaxation through the transfer of adsorbate energy to electron–hole pairs.[7] The number of accessible electron states in the metal increases linearly with the transition frequency, ω_0, and so does the decay rate through

---

6) We assume here that the ground state of the adsorbate is non-decaying. The quantity $\gamma_\perp^s$ is called the transverse relaxation rate. The superscripts s implies that the corresponding quantity is referred to surface.

7) In a metal electron–hole pairs are elementary excitations represented by an electron with energy above the Fermi level and an empty state with energy below this level.

this channel. Therefore, in the case of vibrational relaxation, electron–hole pair excitation is often a dominant pathway for high-frequency vibrational modes, well above the Debye frequency. The coupling between the adsorbate and the electronic density of states in the substrate can be enhanced by a dynamic charge transfer which takes place when the level of a vibrationally excited molecule is close to the Fermi level. In the course of vibration, the occupation of the molecular level oscillates at the vibrational frequency. This effect increases the dynamic dipole moment of the molecule and hence the rate of energy transfer. Depending on the adsorbate–metal system, the decay rates of vibrationally excited molecules via electron–hole pairs range from 10^9 to 10^{12} s^{-1}.

A similar decay mechanism is also possible for adsorbates on semiconductor surfaces. If the transition energy $\hbar\omega_0$ exceeds the forbidden energy gap, which typically ranges from fractions of an eV to a few eV, the adsorbate excitation can be removed through the transition of an electron from the valence band to the conduction band, i.e., via the creation of an electron–hole pair.

Decaying image
The coupling between adsorbate and conduction electrons or optical phonons in the substrate can be treated in the framework of image theory. The damping mechanism is associated with the phase shift between the original dipole field and the one reflected from the surface. In the case of a nonperfect conductor the out-of-phase part of the field is described by the imaginary part of the substrate dielectric function. The losses through coupling with conduction electrons can be understood as energy transfered into Joule heat from the current induced in the substrate by an oscillating dipole.

In a quantum-electrodynamical approach the general expression for the substrate contribution to the rate of the relaxation transition from the excited state $|a\rangle$ follows from Fermi's golden rule and has the form (Wylie and Sipe 1985)

$$R_a^s = \frac{2}{\hbar} \sum_{nj} \mu_j^{an} \mu_j^{na} \, \text{Im} \, G_{jj}(\omega_{na}) \Theta(\omega_{an}) \tag{2.135}$$

where all the notations are the same as in Eq. (2.101). The summation in Eq. (2.135) runs over all quantum states below the given state $|a\rangle$ connected with it by dipole-allowed transitions. If an atom is located above a perfect conductor, the functions G_{jj} are found as (Wylie and Sipe 1984)

$$G_{xx}(i\xi) = G_{yy}(i\xi) = \frac{1}{(8z)^3}(1 + \sigma + \sigma^2)e^{-\sigma} \tag{2.136}$$

$$G_{zz}(i\xi) = \frac{1}{(4z)^3}(1 + \sigma)e^{-\sigma} \tag{2.137}$$

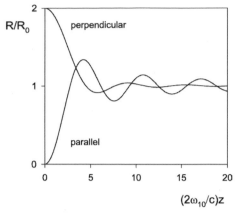

Fig. 2.19 The decay rate of an excited atom as a function of the dimensionless distance from the surface of a perfect conductor, $\eta = 2\omega_{10}z/c$. The two curves correspond to different orientations of the transition dipole moment with respect to the surface.

with $\sigma = 2\xi z/c$. In such a case, the transition rate from the first excited state $|1\rangle$ to the ground state $|0\rangle$ is given by

$$\frac{R_\perp}{R_0} = 1 + 3\left[\frac{\sin(\eta)}{\eta^3} - \frac{\cos(\eta)}{\eta^2}\right] \tag{2.138}$$

or

$$\frac{R_\parallel}{R_0} = 1 - \frac{3}{2}\left[\frac{\cos(\eta)}{\eta^2} + \left(\frac{1}{\eta} - \frac{1}{\eta^3}\right)\sin(\eta)\right] \tag{2.139}$$

with $\eta = 2\omega_{10}z/c$ and R_0 the decay rate of an atom in free space, depending on whether the transition dipole moment is perpendicular or parallel to the surface.

Figure 2.19 shows the dependence of the decay rate corresponding to this mechanism on distance from the surface. The oscillations for $z \geq \lambda$, with λ the wavelength of the transition, arise from the interference between the electromagnetic field emitted by the atom and that reflected from the surface. At shorter distances nonradiative transfer of the excitation energy to the substrate comes into play. In the limit $z \to 0$ one finds (see Eqs (2.138) and (2.139)) that $R_\perp = 2R_0$ with R_0 the decay rate far from the surface and $R_\parallel = 0$. This result is expected from image theory (Section 2.2.1). In the case of a perfect conductor ($\epsilon \to \infty$) the image of the dipole perpendicular to the surface is equal and parallel to it, i.e., the resulting dipole field is twice its value in free space. In contrast, the image of the dipole parallel to the surface is antiparallel to it. This leads to exact cancellation of the field.

In the short-distance limit, $z \ll \lambda$, the decay rates are given by

$$R_\perp^s = \frac{|\mu_\perp^{10}|^2}{2\hbar z^3}\,\mathrm{Im}\left[\frac{\epsilon(\omega_{01}) - 1}{\epsilon(\omega_{01}) + 1}\right] \tag{2.140}$$

and

$$R_{\parallel}^s = \frac{|\mu_{\parallel}^{10}|^2}{4\hbar z^3} \, \text{Im} \left[\frac{\epsilon(\omega_{01}) - 1}{\epsilon(\omega_{01}) + 1} \right] \tag{2.141}$$

The relaxation rate in both cases is enhanced if ω_{01} is close to the frequency of elementary surface excitations in the long-wavelength limit, ω_s, which satisfies the equation $\epsilon(\omega_s) + 1 = 0$ (see Sections 2.1.3 and 2.1.4).

2.2.5
Adsorption and Desorption Kinetics

So far we have considered the dynamics of an adsorbate, including a description of the adsorption potential for different adsorption sites and vibrations of adsorbed species in the electronic potential wells. To this picture one should add the kinetic description, i.e., the time evolution of the population of adsorption sites. This involves evaluation of the rates of different processes which influence the coverage of the surface with adsorbates.

We shall begin with the consideration of the *adsorption rate*, i.e., the sticking rate of atoms or molecules approaching the surface from the gas phase. We assume that the gas is in thermodynamical equilibrium. Then the flux density J of gas particles towards the surface is determined by the gas pressure, P, and the gas temperature, T, and is given by

$$J = \frac{P}{\sqrt{2\pi M k_B T}} \tag{2.142}$$

where M is the mass of a gas particle and k_B is Boltzmann's constant. In general, only a portion $S \leq 1$ of the gas particles are stuck on the surface and hence the adsorption rate can be written in the form

$$\frac{dN_s}{dt} = SJ = S \frac{P}{\sqrt{2\pi M k_B T}} \tag{2.143}$$

with N_s the surface number density of adsorbed particles. The quantity S is called the *sticking probability*. In the adsorption model suggested by Langmuir, the adsorption is assumed to occur only at adsorption sites forming a 2D lattice. Denoting the surface number density of the adsorption sites by N_0, one can determine the *surface coverage* $\theta = N_s/N_0$. In Langmuir's model the coverage is assumed to be in the range $0 \leq \theta \leq 1$.

The sticking probability depends on various factors determining the dynamics of adsorption. The adsorption site must be empty to be available for adsorption and therefore the sticking probability is proportional to the fraction of free adsorption sites. For example, in the case of nondissociative adsorption $S(\theta) = \sigma(1 - \theta)$ with σ a proportionality coefficient. In general, S depends on the kinetic energy of the particles arriving at the surface. To be

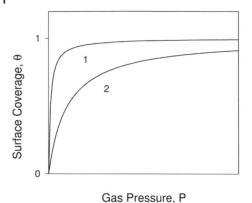

Fig. 2.20 The Langmuir isotherm of adsorption. Curve 1 corresponds to a larger adsorption energy than curve 2.

adsorbed, an atom or a molecule should transfer the energy of translational motion to the substrate which is possible only if the surface can effectively accept this amount of energy. On the other hand, in the course of activated adsorption (Section 2.2.2) a particle should have an incident energy exceeding the activation barrier. In the latter case, the sticking probability obeys an Arrhenius-type temperature dependence

$$S \propto \sigma \propto \exp\left(-\frac{E_a}{k_B T}\right) \tag{2.144}$$

The reverse process to adsorption, i.e., the breaking of the adsorbate–surface bonding and departing of a particle towards the gas phase, is called *desorption*. It is characterized by the desorption rate w which can be represented as

$$w = \nu \exp\left(-\frac{E_d}{k_B T}\right) \tag{2.145}$$

where the pre-exponential factor ν has the dimension of frequency and E_d is the activation energy for desorption.

In equilibrium, the number of particles stuck on the surface per unit time is equal to that desorbed from it. Taking into account Eqs (2.143) and (2.145) we obtain, in the case of nondissociative adsorption, the equation

$$\sigma(1-\theta)J = wN_0\theta \tag{2.146}$$

which determines the equilibrium surface coverage. From here one can deduce the *adsorption isotherm* given by the dependence $\theta(P)$ taken at a constant temperature:

$$\theta(P) = \frac{AP}{1 + AP} \tag{2.147}$$

with

$$A = \frac{1}{\sqrt{2\pi M k_B T}} \frac{\sigma}{w N_0} \qquad (2.148)$$

This so-called *Langmuir isotherm* is shown in Fig. 2.20. For low gas pressures, it can be approximated by a straight line with a slope equal to A. As follows from Eqs (2.148), (2.144), (2.145) and (2.114), its temperature dependence is determined by the binding energy in the adsorption potential:

$$A \propto \frac{1}{\sqrt{T}} \exp\left(\frac{E_b}{k_B T}\right) \qquad (2.149)$$

2.3
Liquid–Solid Interface

Among different liquid–solid interfaces, the boundary between an electrolyte and a metal electrode is the one which has been most investigated in surface science. This is dictated by its importance for electrochemistry and by a rich variety of interesting phenomena. In some respect the relevant processes are similar to those at the gas–solid interface. On the other hand, the electrified character of the electrolyte–solid interface results in some peculiarities. One can control interface properties through external manipulation of the interfacial potential difference. All reactions that involve charge transfer respond directly to this quantity. In this section we shall consider the structure of the electric double layer which takes place at an electrolyte–solid interface and the basic principles of control for various reactions at this boundary.

2.3.1
The Electric Double Layer

Consider the structure of an interface layer between a metal electrode and an electrolyte solution kept under a potential difference ϕ. Due to electrostatic forces, the ions from the solution are attracted by the electrode (*electrostatic adsorption*) and the dipolar molecules are oriented along the lines of the electric force. They can also be physically or chemically adsorbed (*specifically adsorbed*) on the electrode. In the case of electrostatic adsorption alone, ions can approach the electrode to a distance given by their primary solvation shells. The plane parallel to the electrode surface through the centers of electrostatically adsorbed ions at their maximum approach to the electrode is called the *outer Helmholtz plane* (OHP) and the solution region between the OHP and the electrode surface is called the *Helmholtz* or *compact layer*. Due to thermal motion the ions are not confined within the compact layer, but are distributed over the so-called *diffuse layer*. The plane through the centers of specifically adsorbed ions is referred to as the *inner Helmholtz plane* (IHP) (Fig. 2.21).

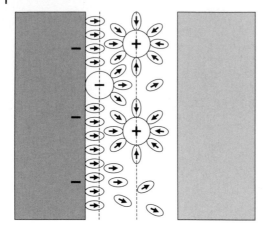

metal   IHP OHP   electrolyte

Fig. 2.21 Structure of the electric double layer.

The electrode–electrolyte interface is thus represented by a double layer, the inner and outer planes of which have opposite charges. In other words, the interface layer has the character of a plate capacitor of molecular size which accumulates a charge. This idea, stemming from Helmholtz, is the basis of most modern concepts. The capacitance of this "capacitor" is huge. Hence very strong electric fields ($\sim 10^7$ V/cm) are obtained in the interface region and very large surface charges (~ 20–$40\,\mu C/cm^2$), corresponding to 0.1–0.2 of the electron charge per surface atom (Kolb 1982).

The sign and the magnitude of the charge accumulated in the compact layer are determined by the applied potential ϕ. Its value at an uncharged electrode is called the *potential of zero charge* (PZC). The dipole moments of the solvent molecules change their orientation with respect to the electrode surface when ϕ scans across the PZC. However, maximum disordering of molecular dipoles occurs at a potential which slightly differs from the PZC. This results from the fact that, besides electrostatic interaction, there is a chemical bonding between the solvent molecules and the electrode surface. This bonding depends on which end of the molecule is attached to the surface. As a result, the orientation of the dipoles does not chang exactly at the PZC.

Figure 2.22 shows the behavior of the electric potential at the electrode–electrolyte interface. The main potential decrease takes place between the electrode surface ($z = 0$) and the outer Helmholtz plane ($z = z_2$). The potential decrease in the diffuse layer, ϕ_2, is much smaller. The overall differential capacitance, $C = \Delta q/\Delta\phi$, of the interface region without specific adsorption can be represented as two capacitances in series, one corresponding to the

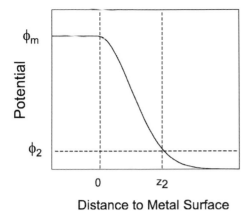

Fig. 2.22 Behavior of the potential across the electric double layer.

compact layer, C_c, and the other to the diffuse layer, C_d:

$$\frac{1}{C} = \frac{1}{C_c} + \frac{1}{C_d} \qquad (2.150)$$

The distance between the plates of the first capacitor is fixed, whereas that of the second one varies as a function of the electrolyte concentration. At low concentrations, the diffuse layer is extended deeply into the solution and the capacitance of the electric double layer is determined mainly by the value of C_d. It has a minimum at $\phi_2 = 0$. Thus the applied potential equals the PZC. This observation can be used for determining the PZC.

2.3.2
Linear-sweep and Cyclic Voltammetry

Variation of the interfacial potential difference, ϕ, allows one to change the charge accumulated in the electric double layer and thus to control reactions which involve charge transfer. For example, for the rate of a reduction process, r, this dependence is given explicitly by (Furtak 1994)

$$r = r_0 \exp\{-[E_a + |e|(\phi - \phi_0)]/k_B T\} \qquad (2.151)$$

where ϕ_0 is the equilibrium potential for the reaction, the pre-exponential factor r_0 contains kinetic parameters and the surface concentrations of the reagents, and E_a is the activation energy. By convention, ϕ is the potential of the sample (working) electrode with respect to a second, reference electrode, at which a well-defined charge transfer equilibrium can be maintained.

The potential of the working electrode can be ramped at a fixed scan rate v. The resulting current density measured as a function of the applied potential is called a *voltammogram*. In *linear-sweep voltammetry*, the potential of the

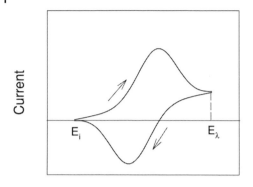

Electrode Potential

Fig. 2.23 Schematic cyclic voltammogram for the reduction reaction at an electrode.

working electrode is ramped from an initial potential E_i to a final potential E_f. *Cyclic voltammetry* can be represented as two subsequent linear-sweep voltammetries: first, the potential is ramped from E_i but at the *switch potential E_λ*, the direction of the potential scan is reversed and then stops at the initial value E_i. The corresponding current–voltage curve makes a loop which is termed the *cyclic voltammogram* (CV).

A typical CV contains peaks which provide information about reactions taking place at the electrode during the scan (Fig. 2.23). These peaks are of similar shape in both forward and reverse scan directions and, in the case of fully reversible reactions, they have identical magnitudes (Monk 2001).

2.4
Gas–Solid Interface

In Section 2.2.5 we have considered adsorption and desorption, the processes which occur at a gas–solid interface and determine the surface coverage with an adsorbate. The same phenomena dictate the gas properties in the close vicinity of the surface. With increasing distance from the surface their influence vanishes due to intermolecular collisions in the gas phase. Therefore, one can define the *gas boundary layer* (GBL) as a gas slab where some properties are different from those in the gas interior. The higher the gas pressure, the more frequent are intermolecular collisions. Accordingly, the thickness of the GBL varies inversely with the pressure.

Any optical signal from a gas is determined by an average over the ensemble of gas molecules and hence by their velocity distribution functions. In this section we shall consider kinetic properties of the GBL in an ideal monatomic

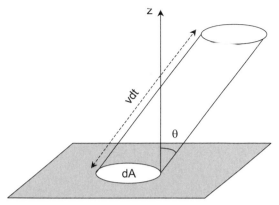

Fig. 2.24 A slanted cylinder illustrating the calculation of the flux distribution function. θ is the polar angle which determines the direction of the velocity vector **v** relative to the surface normal.

gas, both for the equilibrium and nonequilibrium cases. We shall introduce the basic quantities which determine its optical response.

2.4.1
Velocity Distribution Function

An equilibrium gas is characterized by the Maxwellian velocity distribution function

$$f_M(\mathbf{v}) = \frac{1}{\pi^{3/2}v_T^3} \exp\left(-\frac{v^2}{v_T^2}\right) \tag{2.152}$$

where

$$v_T = \sqrt{\frac{2k_BT}{M}} \tag{2.153}$$

is the most probable thermal velocity with k_B the Boltzmann constant, T the gas temperature and M the mass of the gas atom. This means that the quantity $f_M(\mathbf{v})d\mathbf{v}$ gives the fraction of gas atoms whose velocities $\mathbf{v} = (v_x, v_y, v_z)$ lie in the range between $\mathbf{v}$ and $\mathbf{v} + d\mathbf{v}$. In gas–surface collisions, the distribution function (2.152), which refers to the gas as a whole, is of secondary importance. More important are the kinetic properties of gas atoms which strike a surface (Goodman and Wachman 1976). The corresponding velocity distribution function can be found as follows.

Let us consider a slanted cylinder of length vdt, constructed on a small surface area dA and containing a large number of gas atoms (Fig. 2.24).

During the time dt all atoms having velocities in the interval between $\mathbf{v}$ and $\mathbf{v} + d\mathbf{v}$ and confined in the cylinder will strike the surface.[8] Their number is given by

$$dN = nv\cos\theta f_M(\mathbf{v})d\mathbf{v}dAdt \qquad (2.154)$$

where n is the number density of gas atoms and θ is the angle between the velocity vector $\mathbf{v}$ and the surface normal. The total number of atoms impinging the surface area dA for a time dt is found by the integration of Eq. (2.154) over $\mathbf{v}$ and has the form

$$N = \frac{1}{2\sqrt{\pi}}nv_T dAdt \qquad (2.155)$$

Therefore, the fraction of atoms which strikes the surface with a velocity between $\mathbf{v}$ and $\mathbf{v} + d\mathbf{v}$ is obtained as

$$\frac{dN}{N} = f_K(\mathbf{v})d\mathbf{v} \qquad (2.156)$$

where we have introduced the velocity distribution function of atoms impinging the surface,

$$f_K(\mathbf{v}) = 2\sqrt{\pi}\frac{v_z}{v_T}f_M(\mathbf{v}) = \frac{2}{\pi v_T^4}v_z\exp\left(-\frac{v^2}{v_T^2}\right) \qquad (2.157)$$

with $v_z = v\cos\theta$.

The principle of detailed balance requires that, in thermal equilibrium, the velocity distribution of gas atoms leaving the surface is identical with that of gas atoms arriving at the surface, except that v_z changes its sign. Hence, the velocity distribution function of the total atomic flux departing from the surface is also given by Eq. (2.157). It is characterized by a $\cos\theta$-distribution over the polar angles (Fig. 2.25). This flux is strongly anisotropic: it has its maximum along the normal to the surface and zero intensity along the surface. An attempt to generalize this result to nonequilibrium conditions is known as Knudsen's *cosine law*. However, generally in nonequilibrium desorption the angular distribution is described by $\cos^n\theta$ with $0.6 \leq n \leq 12$ (Comsa and David 1985). In such a case, a more detailed consideration of the gas–surface scattering process is necessary.

2.4.2
Scattering Kernel

The Maxwellian velocity distribution function (2.152) represents a steady-state solution of the Boltzmann equation describing the microscopic evolution of a

---

8) When considering equilibrium gas properties, one can ignore collisions between gas atoms, because they do not change the velocity distribution function.

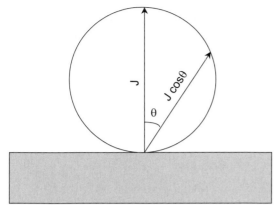

Fig. 2.25 The cosine law for the atomic flux departing from the surface. The intensity of the departing flux is proportional to the length of the chord drawn from the bottom point of the circle.

gas in its interior. To take into account the influence of the border, this equation must be accompanied by a boundary condition at the surface. Quite generally it can be written in the form (Lifshitz and Pitaevskii 1986; Cercignani 2000)

$$f^+(\mathbf{v}') = \int R(\mathbf{v} \to \mathbf{v}')f^-(\mathbf{v})d\mathbf{v} \qquad (2.158)$$

where the velocity distribution functions $f^-(\mathbf{v})$ (with $v_z < 0$) and $f^+(\mathbf{v}')$ (with $v_z' > 0$) describe the atomic fluxes arriving at the surface and departing from it, respectively. The quantity $R(\mathbf{v} \to \mathbf{v}')$ is called the *gas–surface scattering kernel*. It has the sense of a probability density for an atom to strike the surface with velocity between $\mathbf{v}$ and $\mathbf{v}+d\mathbf{v}$ and then to scattered back into the gas with velocity between $\mathbf{v}'$ and $\mathbf{v}' + d\mathbf{v}'$. In thermodynamic equilibrium between the gas and the surface both atomic fluxes have velocity distribution functions given by Eq. (2.157) and hence the function $R(\mathbf{v} \to \mathbf{v}')$ must satisfy the equation

$$f_K(\mathbf{v}') = \int R(\mathbf{v} \to \mathbf{v}')f_K(\mathbf{v})d\mathbf{v} \qquad (2.159)$$

A simple model for gas–surface scattering, proposed by Maxwell, can be represented by the kernel

$$R(\mathbf{v} \to \mathbf{v}') = Sf_K(\mathbf{v}') + (1 - S)\delta(\mathbf{v}'_\| - \mathbf{v}_\|)\delta(v'_z + v_z) \qquad (2.160)$$

where $\mathbf{v}_\|$ is the projection of the vector $\mathbf{v}$ onto the surface plane, $\delta(x)$ is Dirac's delta function and S is the sticking probability (see Section 2.2.5). This model implies that a fraction S of atoms striking the surface gets adsorbed and then desorbs, being in equilibrium with the surface, according to the cosine law. The remaining fraction $(1 - S)$ (directly scattered fraction) undergoes a reflection in the specular direction.

In a more sophisticated model, the reflection from the surface can be mediated by substrate phonons. This results in a broadening of the angular distribution of the reflected atoms. Besides this, one has to take into account the dependence of the sticking probability on the incident energy of the atoms. We thus arrive at the scattering kernel in the form:

$$R(\mathbf{v} \to \mathbf{v}') = S(\mathbf{v})f_K(\mathbf{v}') + CN(\epsilon, \mathbf{Q}) \tag{2.161}$$

where $N(\epsilon, \mathbf{Q})$ is the probability that the state of the phonon system after scattering differs from its initial state by energy ϵ and momentum $\mathbf{Q}$, which are related to the velocities $\mathbf{v}$ and $\mathbf{v}'$ by the conservation laws. C is a constant to be determined from the normalization condition (Brako and Newns 1982).

The scattering kernel is a fundamental quantity summarizing the basic information on the gas–surface interaction. It is indirectly contained in the velocity distribution function of the atomic flux departing from the surface (Eq. (2.158)). Therefore, the atoms "remember" how they have been scattered by the surface as long as interatomic collisions in the gas do not destroy this information. This destruction happens at distances of the order of the mean free path from the surface. The corresponding gas slab nearby the surface is called the *Knudsen layer*.

2.4.3
Relaxation of Gas Excitation

A gas atom which is in an excited state has a finite lifetime due to the possibility of decay to the lower levels. Also, its induced dipole moment can undergo random perturbations disrupting its phase. If an atom can be modelled by a two-level system, both the lifetime of the excited state, T_1, and the dephasing time, T_2', contribute to the optical linewidth $\gamma_\perp = 2/T_2$ according to Eq. (2.134). For a pure radiative decay $T_2 = 2T_1$. Collisions in the gas provide an additional relaxation channel. At moderate gas pressures, the collisional broadening near the atomic transition frequency is proportional to the number density of gas atoms, n, and is determined as

$$\gamma_c = 2n\langle u\sigma' \rangle \tag{2.162}$$

where u is the relative velocity of colliding particles, σ' is the effective cross-section of broadening and the angular brackets denote averaging over velocities. In this case, the total linewidth is given by the sum of radiative and collisional contributions.

The collisional broadening in a monatomic gas is characterized by values of σ' of the order of $(1\text{–}5) \times 10^{-12}$ cm^2. If the perturbing particles are atoms or molecules of a foreign gas, then typically $\sigma' \sim 10^{-14}\text{–}10^{-13}$ cm^2. A similar mechanism of broadening occurs also for vibrational and rotational transitions in molecules. The corresponding cross-sections are much less, about $(1\text{–}3) \times 10^{-15}$ cm^2.

Close to the surface the relaxation mechanisms considered in Section 2.2.4 become prominent. Usually, the relevant relaxation times are much shorter than those far away from the surface. An atom which gets adsorbed on the surface for a time longer than the relaxation times T_1^s and T_2^s will be desorbed being in its ground state with completely quenched induced dipole moment. As a result, the atomic state populations as well as the polarization in the atomic flux departing from the surface are different from those in the flux arriving at it. This distinction carries information on the relaxation processes in the close vicinity of the surface. It is reasonable to introduce "memory lengths" which determine the characteristic distances at which this information is still contained in the departing flux. Then the *population memory length* can be defined as $l_1 = v_T T_1$, whereas the *polarization memory length* $l_2 = v_T T_2$ (Bordo and Rubahn 2004).

Further Reading

S.G. Davison, M. Stęślicka, *Basic Theory of Surface States*, Oxford University Press, Oxford, 1992.

J.W. Gadzuk, A.C. Luntz, On vibrational lineshapes of adsorbed molecules, *Surf. Sci.* **1984**, *144*, 429.

F.O. Goodman, H.Y. Wachman, *Dynamics of Gas-Surface Scattering*, Academic Press, New York, 1976.

T.B. Grimley, Theory of Chemisorption in *The Chemical Physics of Solid Surfaces and Heterogeneous Catalysis*, D.A. King, D.P. Woodruff (Eds.), Elsevier, Amsterdam, 1983.

J. Koryta, J. Dvořák, L. Kavan, *Principles of Electrochemistry*, John Wiley & Sons, Chichester, 1993.

H. Lüth, *Surfaces and Interfaces of Solid Materials*, Springer-Verlag, Berlin, 1995.

R.G. Tobin, Vibrational linewidths of adsorbed molecules: Experimental considerations and results, *Surf. Sci.* **1987**, *183*, 226.

R.F. Willis, A.A. Lucas, G.D. Mahan, Vibrational Properties of Adsorbed Molecules in *The Chemical Physics of Solid Surfaces and Heterogeneous Catalysis*, D.A. King, D.P. Woodruff (Eds.), Elsevier, Amsterdam, 1983.

A. Zangwill, *Physics at Surfaces*, Cambridge University Press, Cambridge, 1988.

Problems

Problem 2.1. Draw the arrangement of atoms in the (111) surface of: (a) a bulk centered cubic (bcc) crystal; (b) a face centered cubic (fcc) crystal. Show the surface unit cell and indicate the $[\bar{1}10]$ and $[\bar{1}\bar{1}2]$ axes in both cases. What symmetry do the surface Bravais lattices have?

Problem 2.2. Figure 2.26 represents the electronic structure of the Cu(111) surface. Show the electronic surface states. What does the parabolic shape of the upper surface band mean? How can one classify this state? Which state can one refer to as a surface resonance? Estimate the minimum energy of the transition between the surface states.

Problem 2.3. Figure 2.27 represents the dispersion of phonons on the NaF(100) surface. Show the dispersion curves corresponding to surface phonons. Which of them are related to: (a) acoustical phonon branches; (b) optical phonon branches? Which of them can be referred to as surface resonance phonons?

Problem 2.4. Find the power spectrum of a corrugated surface described by the profile

$$\zeta(x,y) = h \cos\left(\frac{2\pi}{a}x\right) \cos\left(\frac{2\pi}{b}y\right) \tag{2.163}$$

Note: Use an equation analogous to Eq. (2.32).

Problem 2.5. Estimate the distance from a glass surface at which the frequency shift of the $3S_{1/2} \to 3P_{3/2}$ transition in sodium atoms becomes comparable with its natural linewidth ($\gamma = 2\pi \times 10$ MHz). Assume that an atom can be modelled by a harmonic oscillator and, hence, its polarization as well as the van der Waals shift are independent of the state. The index of refraction of glass at the transition frequency is $n = 1.5$ and the average transition dipole moment of Na atoms is $\mu_{01} = 5.2$ D.[9]

Problem 2.6. How many different vibrational frequencies does an atom have when being adsorbed on a (111) surface of a cubic crystal: (a) at a hollow site; (b) at an on-top site; (c) at a bridge site?

Problem 2.7. Estimate the distance between a sodium atom and a gold surface at which the surface-induced relaxation rate at the $3S_{1/2} \to 3P_{3/2}$ atomic transition is comparable with that due to the natural decay rate. Assume that the nonradiative relaxation can be described by a decaying image. The natural lifetime of the $3P_{3/2}$ state is $\tau = 16.3$ ns, the average transition dipole moment

9) 1 D (Debye) $= 10^{-18}$ cgse.

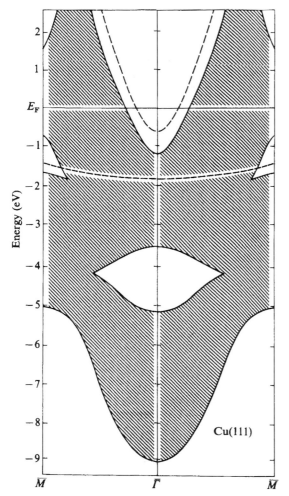

Fig. 2.26 Calculated electronic structure of Cu(111) surface (Euceda et al. 1983). E_F is the Fermi energy. Reprinted with permission from (Zangwill 1988). Copyright 1988, Cambridge University Press.

of an Na atom is $\mu_{01} = 5.2$ D and the complex index of refraction of gold at the transition frequency is $\tilde{n} = 0.29 + i \cdot 2.86$.

Problem 2.8. The flux of sodium atoms to the glass surface having a temperature $T = 300$ K is equal to $J = 5 \cdot 10^{14}$ cm^{-2}s^{-1}. Estimate the steady-state surface coverage with Na adatoms assuming the Langmuir model of adsorption. Take for the estimate the pre-exponential factor for the rate of desorption $v = 2 \cdot 10^{17}$ s^{-1}, the binding energy of adatoms $E_b = 0.8$ eV, the surface density of adsorption centers $N_0 = 10^{15}$ cm^{-2} and the sticking probability at a free adsorption center, $\sigma = 1$.

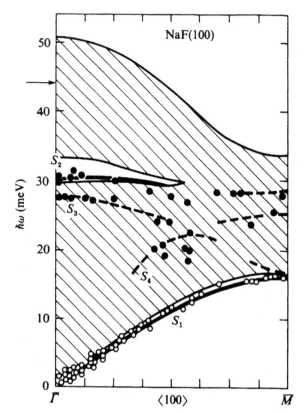

Fig. 2.27 Calculated dispersion of phonons on the NaF(100) surface. Experimental data are shown by filled and open circles (Brusdeylins et al. 1985). Reprinted with permission from (Zangwill 1988). Copyright 1988, Cambridge University Press.

Problem 2.9. The atomic flux desorbing from the surface obeys the Knudsen law. Find the most probable velocity component normal to the surface in terms of the most probable thermal velocity in a gas volume, v_T.

3
Linear Optical Properties of Surfaces and Interfaces

In this chapter we begin to study optical phenomena which occur at surfaces
or interfaces when the intensity of the incident light is weak. These phenomena include reflection, refraction and the scattering of light, as well as the excitation of coupled modes propagating along the surface or interface. The
optical properties of media at the border, which determine these phenomena,
are assumed to be independent of the light intensity. Nonlinear optical effects
will be discussed in Chapter 6.

The discussion of optical properties is quite general as far as it is given in
terms of the dielectric functions of the contacting media. Hence, it is equally
valid for solids, liquids and gases, except for some explicitly specified cases.
The special case of optical response at a gas–solid interface near a resonance
frequency of the gas is discussed later in Chapter 7.

3.1
Reflection and Refraction of Light

The most prominent optical properties of surfaces and interfaces which we
face in our everyday life are represented by the reflection and refraction of
light. These phenomena occur at all boundaries between different media and
due to them we can identify objects or are able to use various optical devices.
The human sight itself is possible due to refraction of light in the eye. The
laws that govern these phenomena were first derived by Fresnel in 1823.

3.1.1
Fresnel Equations

Let us consider a plane electromagnetic wave

$$\mathbf{E}_i(\mathbf{r}, t) = \mathbf{E}_{i0}e^{i\mathbf{k}_{i\|}\cdot\mathbf{r}_\|}e^{ik_{iz}z}e^{-i\omega t} \tag{3.1}$$

Optics and Spectroscopy at Surfaces and Interfaces. Vladimir G. Bordo and Horst-Günter Rubahn
Copyright © 2005 WILEY-VCH Verlag GmbH & Co. KGaA, Weinheim
ISBN: 3-527-40560-7

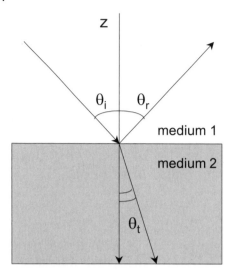

Fig. 3.1 Geometry of reflection and refraction of light.

incident from medium 1 onto a plane boundary with medium 2 (Fig. 3.1). The z-axis is chosen along the normal to the boundary towards medium 2. Both media are assumed to be homogeneous and isotropic and are described by the dielectric functions ϵ_1 and ϵ_2, the latter being in general a complex quantity. The wave is split into two waves: a transmitted fraction, which propagates into the second medium; and a fraction, which is reflected back into the first medium. Since we assume complete homogeneity along the boundary plane, the dependence of the wave amplitudes on $\mathbf{r}_\|$ must be the same for all three waves. This gives

$$\mathbf{k}_{i\|} = \mathbf{k}_{r\|} = \mathbf{k}_{t\|} \tag{3.2}$$

where the subscripts r and t refer to the reflected and transmitted waves, respectively. The moduli of the wave vectors in media 1 and 2 are given by[1]

$$k_i = k_r = \frac{\omega}{c}\sqrt{\epsilon_1} = \frac{\omega}{c}n_1 \tag{3.3}$$

and

$$k_t = \frac{\omega}{c}\sqrt{\epsilon_2} = \frac{\omega}{c}n_2 \tag{3.4}$$

where c is the speed of light, and n_1 and n_2 are the refractive indices of the corresponding media. The components of the wave vectors parallel to the surface are found as (see Eq. (3.2))

$$k_{i\|} = \frac{\omega}{c}\sqrt{\epsilon_1}\sin\theta_i = \frac{\omega}{c}\sqrt{\epsilon_1}\sin\theta_r = \frac{\omega}{c}\sqrt{\epsilon_2}\sin\theta_t \tag{3.5}$$

1) Here and in the following we assume that $\mu = 1$ in both media.

and the z-components of the wave vectors have the forms

$$k_{rz} = -k_{iz} = -\frac{\omega}{c}\sqrt{\epsilon_1}\cos\theta_i \qquad (3.6)$$

and

$$k_{tz} = \frac{\omega}{c}\sqrt{\epsilon_2 - \epsilon_1\sin^2\theta_i} \qquad (3.7)$$

where θ_i, θ_r and θ_t are the angles which the corresponding waves make with the normal to the boundary. From this the laws of reflection and refraction (Snell's laws) follow:

$$\theta_r = \theta_i \qquad (3.8)$$

$$\frac{\sin\theta_t}{\sin\theta_i} = \sqrt{\frac{\epsilon_1}{\epsilon_2}} = \frac{n_1}{n_2} \qquad (3.9)$$

The amplitudes of the reflected and transmitted waves, $\mathbf{E}_r$ and $\mathbf{E}_t$, are found from the boundary conditions to Maxwell's equations. We shall consider the two independent polarizations of the incident wave separately (Landau and Lifshitz 1963; Born and Wolf 1975).

s-polarization
If $\mathbf{E}_{i0}$ is perpendicular to the plane of incidence containing the $\mathbf{k}_i$ vector and the normal to the boundary, the polarization state is called s-*polarized*. The symmetry of the problem with respect to this plane indicates that the same polarization is obtained for the amplitudes of the reflected and refracted waves. With the y-axis along the $\mathbf{E}_{i0}$ vector, we can write the boundary conditions at $z = 0$ as

$$E_{i0y} + E_{r0y} = E_{t0y} \qquad (3.10)$$

$$H_{i0x} + H_{r0x} = H_{t0x} \qquad (3.11)$$

where $\mathbf{H}_{\alpha 0}$ ($\alpha = i, r, t$) are the amplitudes of the magnetic field for the corresponding waves. The latter quantities can be expressed in terms of $\mathbf{E}_{\alpha 0}$ using the relation $H_x = -ck_z E_y/\omega$ which follows from the Maxwell equation $\nabla \times \mathbf{E} = -(1/c)(\partial\mathbf{H}/\partial t)$. The exponential factors on both sides of Eqs (3.10) and (3.11) have been cancelled due to Eq. (3.2). As a result, one obtains

$$E_{r0} = \frac{\sqrt{\epsilon_1}\cos\theta_i - \sqrt{\epsilon_2 - \epsilon_1\sin^2\theta_i}}{\sqrt{\epsilon_1}\cos\theta_i + \sqrt{\epsilon_2 - \epsilon_1\sin^2\theta_i}}E_{i0} \qquad (3.12)$$

$$E_{t0} = \frac{2\sqrt{\epsilon_1}\cos\theta_i}{\sqrt{\epsilon_1}\cos\theta_i + \sqrt{\epsilon_2 - \epsilon_1\sin^2\theta_i}}E_{i0} \qquad (3.13)$$

Analogous equations can be written for the amplitudes of the magnetic field (see Landau and Lifshitz (1963) for details).

The *reflection coefficient, r*, is defined as the ratio of reflected to incident electric field amplitude

$$r = \frac{E_{r0}}{E_{i0}} \qquad (3.14)$$

The *reflectivity, R*, is defined as the ratio of the time averaged z-component of the Poynting vector of the reflected wave to that of the incident wave. The *transmissivity, T*, is defined as a similar quantity for the transmitted wave. The Poynting vector component of the corresponding wave normal to the boundary is given by

$$\bar{S}_z = \frac{c\sqrt{\epsilon_{1(2)}}}{8\pi} |\mathbf{E}_{\alpha 0}|^2 \cos\theta_\alpha = \frac{c}{8\pi\sqrt{\epsilon_{1(2)}}} |\mathbf{H}_{\alpha 0}|^2 \cos\theta_\alpha \qquad (3.15)$$

From here one finds

$$R_s = |r_s|^2 = \left| \frac{\sqrt{\epsilon_1}\cos\theta_i - \sqrt{\epsilon_2 - \epsilon_1\sin^2\theta_i}}{\sqrt{\epsilon_1}\cos\theta_i + \sqrt{\epsilon_2 - \epsilon_1\sin^2\theta_i}} \right|^2 \qquad (3.16)$$

and

$$T_s = \frac{4\left|\sqrt{\epsilon_1}\cos\theta_i\sqrt{\epsilon_2 - \epsilon_1\sin^2\theta_i}\right|}{\left|\sqrt{\epsilon_1}\cos\theta_i + \sqrt{\epsilon_2 - \epsilon_1\sin^2\theta_i}\right|^2} \qquad (3.17)$$

where the subscript s indicates the polarization of the incident wave and we have used Eq. (3.9).

p-polarization

Let us now describe the case of p-*polarization* where the vector $\mathbf{E}_{i0}$ lies in the plane of incidence. The calculations can be carried out much as for the s-polarization. This time the magnetic field is perpendicular to the plane of incidence and choosing the y-axis along it we can write the boundary conditions as

$$H_{i0y} + H_{r0y} = H_{t0y} \qquad (3.18)$$

$$E_{i0x} + E_{r0x} = E_{t0x} \qquad (3.19)$$

The amplitudes $E_{\alpha 0x}$ can be expressed in terms of $H_{\alpha 0y}$ using the relation $E_x = ck_z H_y/\epsilon\omega$ which follows from Maxwell's equation $\nabla \times \mathbf{H} =$

$(\epsilon/c)(\partial \mathbf{E}/\partial t)$. The solution of Eqs (3.18) and (3.19) is given by

$$H_{r0} = \frac{\epsilon_2 \cos \theta_i - \sqrt{\epsilon_1 \epsilon_2 - \epsilon_1^2 \sin^2 \theta_i}}{\epsilon_2 \cos \theta_i + \sqrt{\epsilon_1 \epsilon_2 - \epsilon_1^2 \sin^2 \theta_i}} H_{i0} \qquad (3.20)$$

$$H_{t0} = \frac{2\epsilon_2 \cos \theta_i}{\epsilon_2 \cos \theta_i + \sqrt{\epsilon_1 \epsilon_2 - \epsilon_1^2 \sin^2 \theta_i}} H_{i0} \qquad (3.21)$$

Now using Eq. (3.15) we find

$$R_p = |r_p|^2 = \left| \frac{\epsilon_2 \cos \theta_i - \sqrt{\epsilon_1 \epsilon_2 - \epsilon_1^2 \sin^2 \theta_i}}{\epsilon_2 \cos \theta_i + \sqrt{\epsilon_1 \epsilon_2 - \epsilon_1^2 \sin^2 \theta_i}} \right|^2 \qquad (3.22)$$

and

$$T_p = \frac{4 \left| \epsilon_2 \cos \theta_i \sqrt{\epsilon_1 \epsilon_2 - \epsilon_1^2 \sin^2 \theta_i} \right|}{\left| \epsilon_2 \cos \theta_i + \sqrt{\epsilon_1 \epsilon_2 - \epsilon_1^2 \sin^2 \theta_i} \right|^2} \qquad (3.23)$$

Fresnel Formulae
Equations (3.12) and (3.13), (3.20) and (3.21) are called *Fresnel formulae*. They are valid even when medium 2 is absorbing. For the case when both media are transparent, a more simple form can be obtained using the law of refraction (3.9):

$$E_{r0} = \frac{\sin(\theta_t - \theta_i)}{\sin(\theta_i + \theta_t)} E_{i0} \qquad (3.24)$$

$$E_{t0} = \frac{2 \cos \theta_i \sin \theta_t}{\sin(\theta_i + \theta_t)} E_{i0} \qquad (3.25)$$

$$H_{r0} = \frac{\tan(\theta_i - \theta_t)}{\tan(\theta_i + \theta_t)} H_{i0} \qquad (3.26)$$

$$H_{t0} = \frac{\sin 2\theta_i}{\sin(\theta_i + \theta_t) \cos(\theta_i - \theta_t)} H_{i0} \qquad (3.27)$$

The reflectivities for different polarizations are given by

$$R_s = \frac{\sin^2(\theta_t - \theta_i)}{\sin^2(\theta_i + \theta_t)} \qquad (3.28)$$

and

$$R_p = \frac{\tan^2(\theta_i - \theta_t)}{\tan^2(\theta_i + \theta_t)} \qquad (3.29)$$

They remain the same if one interchanges θ_i and θ_t. This means that the reflectivity for a wave incident from medium 1 at an angle θ_i is equal to that for a wave incident from medium 2 at an angle θ_t.

It should be noted that the denominator in Eq. (3.29) tends to infinity when $\theta_i + \theta_t \to \pi/2$ and hence $R_p \to 0$ under such conditions. Denoting the corresponding incidence angle by θ_B we find from the law of refraction (3.9) that

$$\tan \theta_B = \sqrt{\frac{\epsilon_2}{\epsilon_1}} \tag{3.30}$$

At this angle, independent of the polarization state of the incident light, the reflected light will always be s-polarized. On the other hand, p-polarized light hitting the interface at θ_B will undergo no reflection. The angle θ_B is therefore called the *angle of total polarization* or the *Brewster angle*.

Phase changes

Since the trigonometrical factors on the right-hand sides of Eqs (3.24)–(3.27) are real[2], the phase of each component of the reflected or transmitted wave either remains unchanged or changes by π, depending on the sign of the coefficients. The factors in Eqs (3.25) and (3.27) are positive for all values of θ_i and θ_t and hence the phase of the transmitted wave is equal to that of the incident wave. The change in phase of the reflected wave depends on the relative magnitudes of θ_i and θ_t.

In a more general case where medium 2 is absorbing, i.e., ϵ_2 is a complex quantity, the phase changes on reflection can be found from Eqs (3.12) and (3.20) as

$$\delta_s = \arctan \left[\frac{\mathrm{Im}(r_s)}{\mathrm{Re}(r_s)} \right] \tag{3.31}$$

$$\delta_p = \arctan \left[\frac{\mathrm{Im}(r_p)}{\mathrm{Re}(r_p)} \right] \tag{3.32}$$

In general, $\delta_s \neq \delta_p$, and therefore if the incident linearly polarized wave, has both s- and p-components, the reflected wave is elliptically polarized. The parameters of the polarization ellipse are determined by the optical constants of both media and can be measured with high accuracy. This principle forms the basis of reflection *ellipsometry* (see Section 5.1). For historical reasons one often represents the ratio of the reflection coefficients, ρ, in terms of the ellipsometric angles, Ψ and Δ, as

$$\rho = r_p/r_s = \tan \Psi \exp(i\Delta) \tag{3.33}$$

---

2) We do not consider here the case of total internal reflection.

Skin effect

In the case when $\epsilon_2 < 0$ the quantity k_{tz}, Eq. (3.7), becomes pure imaginary. This means that the wave can penetrate into medium 2 only at the depth of the *skin layer* δ. For normal incidence this quantity is determined as

$$\delta = \frac{\lambda}{2\pi \sqrt{|\epsilon_2|}} \tag{3.34}$$

which is less than the wavelength in vacuum, λ. Such a situation is realized, for example, in metals at frequencies below the plasma frequency (see Eq. (2.64)). For large absolute values of the dielectric function, when $|\epsilon_2| \gg \epsilon_1$, the reflectivity approaches unity, explaining the metallic luster of good conductors.

3.1.2
Total Internal Reflection (TIR)

If we rewrite the law of refraction (3.9) in the form

$$\sin \theta_t = \frac{n_1}{n_2} \sin \theta_i \tag{3.35}$$

we see that for light that is incident from an optically more dense medium (i.e., $n_1 > n_2$), $\sin \theta_t$ becomes equal to 1 (and hence $\theta_t = \pi/2$) if the angle of incidence equals

$$\theta_c = \arcsin \left(\frac{n_2}{n_1} \right) \tag{3.36}$$

When θ_i is further increased, the right-hand side of Eq. (3.35) is greater than unity and θ_t cannot be determined from it. From Eq. (3.7) it follows that the quantity k_{tz} is pure imaginary at such angles, i.e., the wave does not propagate into medium 2. Therefore, all the energy incident on the boundary is reflected back into medium 1, so that $R_s = R_p = 1$. This phenomenon is called *total internal reflection* and the angle θ_c is referred to as the *critical angle*.

In total internal reflection the transmitted wave amplitude has the form

$$\mathbf{E}_t(\mathbf{r}, t) = \mathbf{E}_{t0} e^{-\kappa z} e^{i \mathbf{k}_\parallel \cdot \mathbf{r}_\parallel} e^{-i \omega t} \tag{3.37}$$

where we have introduced the notation

$$\kappa = \operatorname{Im} k_{tz} = \frac{\omega}{c} \sqrt{n_1^2 \sin^2 \theta_i - n_2^2} \tag{3.38}$$

Equation (3.37) describes a wave propagating along the boundary in the plane of incidence whose amplitude decreases exponentially with distance from the boundary. The *penetration depth* δ of this wave into medium 2 at which its amplitude decreases by a factor of e is given by

$$\delta = \frac{1}{\kappa} = \frac{\lambda}{2\pi} \frac{1}{\sqrt{n_1^2 \sin^2 \theta_i - n_2^2}} \tag{3.39}$$

This quantity has the order of magnitude $\sim \lambda$ for incidence angles far from the critical angle, i.e., the wave has only a vanishing (or *evanescent*) tail penetrating into medium 2. For this reason the wave (3.37) which appears in total internal reflection is called an *evanescent wave* (EW).

3.1.3
Account of a Transition Layer

In the two preceding sections we have considered reflection of light for the idealized case of a sharp boundary between two media. However, in practice, one often deals with transition layers of different nature which exist at an interface. Such layers may consist of adsorbed species, surface oxides or may be represented by a thin film deposited on a solid or a liquid surface. Even when the surface can be considered as "clean" a thin boundary region whose optical properties are perturbed by the presence of the surface might be treated as a transition layer.

3.1.3.1 Macroscopic layer
If the layer thickness is much larger than interatomic distances, the interface can be treated as a three-phase system for which equations analogous to the Fresnel formulae can be derived. We shall consider here only the reflection coefficients because they are relevant to different optical techniques (McIntyre and Aspnes 1971).[3]

For the three-phase system (123) shown in Fig. 3.2, the equivalent Fresnel reflection coefficient is defined by analogy to the two-phase system as the ratio of the amplitudes of the reflected and incident waves in the initial ambient phase (1). It can be written for both s- and p-polarizations as

$$r_{123} = \frac{r_{12} + r_{23}e^{2i\beta}}{1 + r_{12}r_{23}e^{2i\beta}} \tag{3.40}$$

where r_{ij} are the reflection coefficients at interfaces between semi-infinite media i and j, and β is the change in phase of the beam on one pass through the layer (phase 2) given by

$$\beta = \frac{2\pi d}{\lambda}\sqrt{\epsilon_2 - c_1 \sin^2 \theta_I} \tag{3.41}$$

with d the layer thickness. All the reflection coefficients in Eq. (3.40) must be

---

3) As distinct from (McIntyre and Aspnes 1971) we assume below, the incident electromagnetic wave in the form (3.1).

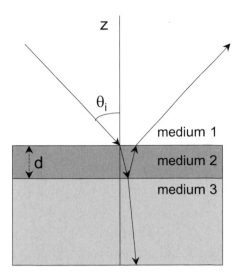

Fig. 3.2 Geometry of reflection of light in a three-phase system.

taken for the same polarization. The corresponding reflectivity is found as

$$R(d) \equiv R_{123} = |r_{123}|^2 \tag{3.42}$$

$$= \frac{R_{12} + R_{23}e^{-4\operatorname{Im}\beta} + 2\sqrt{R_{12}R_{23}}e^{-2\operatorname{Im}\beta}\cos(\delta_{12} - \delta_{23} - 2\operatorname{Re}\beta)}{1 + R_{12}R_{23}e^{-4\operatorname{Im}\beta} + 2\sqrt{R_{12}R_{23}}e^{-2\operatorname{Im}\beta}\cos(\delta_{12} + \delta_{23} + 2\operatorname{Re}\beta)} \tag{3.43}$$

where R_{ij} are the reflectivities at interfaces between semi-infinite media i and j, δ_{ij} are the corresponding phase changes, Eqs (3.31) and (3.32). Equation (3.43) has identical forms for both s- and p-polarizations of the incident light beam.

The three-phase system considered above is reduced to a two-phase one (13) in the limit $d \to 0$. Although in experiments absolute measurements of the reflectivities $R(0)$ and $R(d)$ require much effort, the ratio $R(d)/R(0)$ can be readily measured with high accuracy due to a cancellation of common errors. To facilitate a comparison between experiment and theory, it is worthwhile considering a practically important case where the layer thickness is much less than the wavelength of the incident radiation. In such a case, expression (3.40) can be expanded to terms of first order in β. In this approximation, the reflection coefficients normalized to their values at an uncovered surface can be written as

$$\frac{r_s(d)}{r_s(0)} = 1 + \frac{4\pi i d n_1 \cos\theta_i}{\lambda}\left(\frac{\epsilon_2 - \epsilon_3}{\epsilon_1 - \epsilon_3}\right) \tag{3.44}$$

and

$$\frac{r_p(d)}{r_p(0)} = 1 + \frac{4\pi i d n_1 \cos\theta_i}{\lambda}\left\{\frac{\epsilon_2 - \epsilon_3}{\epsilon_1 - \epsilon_3}\left[\frac{1 - (\epsilon_1/\epsilon_2\epsilon_3)(\epsilon_2 + \epsilon_3)\sin^2\theta_i}{1 - (1/\epsilon_3)(\epsilon_1 + \epsilon_3)\sin^2\theta_i}\right]\right\} \tag{3.45}$$

for s- and p-polarization, respectively. The corresponding normalized reflectivity changes,

$$\frac{\Delta R}{R} = \frac{R(d) - R(0)}{R(0)} \tag{3.46}$$

are obtained as

$$\frac{\Delta R_s}{R_s} = -\frac{8\pi dn_1 \cos\theta_i}{\lambda} \operatorname{Im}\left(\frac{\epsilon_2 - \epsilon_3}{\epsilon_1 - \epsilon_3}\right) \tag{3.47}$$

and

$$\frac{\Delta R_p}{R_p} = -\frac{8\pi dn_1 \cos\theta_i}{\lambda} \operatorname{Im}\left\{\frac{\epsilon_2 - \epsilon_3}{\epsilon_1 - \epsilon_3}\left[\frac{1 - (\epsilon_1/\epsilon_2\epsilon_3)(\epsilon_2 + \epsilon_3)\sin^2\theta_i}{1 - (1/\epsilon_3)(\epsilon_1 + \epsilon_3)\sin^2\theta_i}\right]\right\} \tag{3.48}$$

At normal incidence the result for both polarizations is given by

$$\frac{\Delta R}{R} = -\frac{8\pi dn_1}{\lambda} \operatorname{Im}\left(\frac{\epsilon_2 - \epsilon_3}{\epsilon_1 - \epsilon_3}\right) \tag{3.49}$$

The relations (3.47) and (3.48) are equivalent to the ones originally derived by Drude (1891). One can see from here that if the dielectric functions of the media vary slowly with frequency, the sensitivity for film detection is linearly proportional to the film thickness and inversely proportional to the wavelength. From Eqs (3.47) and (3.48) it also follows that $\Delta R/R = 0$, if ϵ_1, ϵ_2 and ϵ_3 are all real. This result occurs only to first order in d/λ, whereas interference effects arise to second order, manifesting themselves as an enhancement or a suppression of the reflectivity, depending on the ratio between d and λ.

If the transition layer is anisotropic with the dielectric tensor components ϵ_{jj}, $j = x, y, z$, and the ambient phase (1) is a vacuum, Eqs (3.44) and (3.45) are modified to (Hingerl et al. 1993)

$$\frac{r_{ss}(d)}{r_{ss}(0)} = 1 + \frac{4\pi id\cos\theta_i}{\lambda}\left(\frac{\epsilon_b - \bar{\epsilon} - \Delta\epsilon\cos 2\psi}{\epsilon_b - 1}\right) \tag{3.50}$$

and

$$\frac{r_{pp}(d)}{r_{pp}(0)} = 1 + \frac{4\pi id\cos\theta_i}{\lambda(\epsilon_b - 1)(\epsilon_b\cos^2\theta_i - \sin^2\theta_i)}\left\{(\epsilon_b - \bar{\epsilon})\epsilon_b - [(\epsilon_b^2/\epsilon_{zz}) - \bar{\epsilon}]\sin^2\theta_i\right.$$
$$\left. + \Delta\epsilon(\epsilon_b - \sin^2\theta_i)\cos 2\psi\right\} \tag{3.51}$$

where $\bar{\epsilon} = (\epsilon_{xx} + \epsilon_{yy})/2$, $\Delta\epsilon = (\epsilon_{yy} - \epsilon_{xx})/2$, $\epsilon_b \equiv \epsilon_3$ is the isotropic bulk dielectric function and ψ is an azimuthal angle between the plane of incidence and the x-axis of the layer. Here, the first subscript refers to the polarization of reflected light, whereas the second subscript refers to that of incident light. The anisotropy of the layer leads to a mixture of different polarizations in reflection, described by the coefficients

$$r_{ps}(d) = -r_{sp}(d)$$

$$= \frac{4\pi id\cos\theta_i}{\lambda}\frac{\Delta\epsilon\sin 2\psi\sqrt{\epsilon_b - \sin^2\theta_i}}{(\epsilon_b\cos\theta_i + \sqrt{\epsilon_b - \sin^2\theta_i})(\cos\theta_i + \sqrt{\epsilon_b - \sin^2\theta_i})} \tag{3.52}$$

It should be noted that Eqs (3.47) and (3.48) are valid also for the case of total internal reflection and can be applied to fractionally covered surfaces through the use of a mean film thickness

$$\bar{d} = \frac{1}{A} \sum_i d_i \delta A_i \tag{3.53}$$

where d_i and δA_i are the thickness and the area of the ith island, and A is the area of the bare substrate surface. If the film consists of uniform islands of equal heights d, this quantity can be written as

$$\bar{d} = d\theta \tag{3.54}$$

with

$$\theta = \frac{\sum_i \delta A_i}{A} \tag{3.55}$$

the surface coverage.

3.1.3.2 Microscopic layer

In the case where the transition layer is represented by adsorbed atoms or molecules, or by a perturbed surface layer, its thickness d is of the order of interatomic distances. Such a situation is beyond the applicability limit of the macroscopic Maxwell equations and therefore the above consideration is no longer valid. Instead, one has to take into account the microscopic structure of the transition layer. The first approach to this problem is to consider the polarization of the atomic dipoles, both in the medium and in the surface transition layer, averaged over an infinitesimal volume, as a source of radiation resulting in reflected light. Such an average is carried out to evaluate the radiation field far from the surface and therefore is reasonable for a microscopic layer.

Let a plane electromagnetic wave with frequency ω strike the surface of the medium from the vacuum side. Then the polarization vector of the medium has the form

$$\mathbf{P} = \mathbf{P}_0(z)e^{i(\mathbf{k}_t \cdot \mathbf{r} - \omega t)} \tag{3.56}$$

The transition layer is defined as a layer where the amplitude $\mathbf{P}_0(z)$ varies with z. Each volume element, dV, of both medium and layer, having a time-varying dipole moment $\mathbf{P}dV$, is a source of electromagnetic waves propagating in vacuum with the speed of light. The problem now is to find the total radiation field inside and outside the medium. To do that, one can break up the medium into layers of small thickness Δz by planes parallel to the surface. The polarization amplitude, $\mathbf{P}_0(z)$, within each layer can be considered as a constant. Thus, within the first layer we have a wave of constant amplitude $\mathbf{P}_1 = \mathbf{P}_{01} \exp[i(\mathbf{k}_t \cdot \mathbf{r} - \omega t)]$, within the second layer a wave $\mathbf{P}_2 = \mathbf{P}_{02} \exp[i(\mathbf{k}_t \cdot \mathbf{r} - \omega t)]$, and so on. These waves can be exchanged by the following system of waves with constant amplitudes propagating inside the whole medium but from boundaries of different layers: the first wave, $\mathbf{P}_{01} \exp[i(\mathbf{k}_t \cdot \mathbf{r} - \omega t)]$, propagating from the first plane; the second wave, $(\mathbf{P}_{02} - \mathbf{P}_{01}) \exp[i(\mathbf{k}_t \cdot \mathbf{r} - \omega t)]$, propagating from the second plane; the third wave, $(\mathbf{P}_{03} - \mathbf{P}_{02}) \exp[i(\mathbf{k}_t \cdot \mathbf{r} - \omega t)]$, propagating from the third plane, and so forth.

By this means the problem of evaluating the radiation field originating from the wave (3.56) is reduced to the problem of evaluating the radiation field of a polarization wave with a constant amplitude. This field can be found far from the surface in terms of the Hertz vector which satisfies a nonuniform wave equation with the polarization vector as a source (Born and Wolf 1975).

In this approach the reflection coefficient for s- and p-polarizations in linear approximation in the ratio d/λ can be represented in the form (Sivukhin 1948)

$$\frac{r_s(d)}{r_s(0)} = 1 + \frac{4\pi i}{\lambda} \gamma_y \cos \theta_i \qquad (3.57)$$

and

$$\frac{r_p(d)}{r_p(0)} = 1 + \frac{4\pi i}{\lambda} \cos \theta_i \frac{\gamma_x \cos^2 \theta_t - \gamma_z \sin^2 \theta_i}{\cos^2 \theta_t - \sin^2 \theta_i} \qquad (3.58)$$

Here, the quantities γ_j ($j = x, y, z$) can be written as

$$\gamma_j = d \frac{\Delta P_j}{P_{3j}} \qquad (3.59)$$

with

$$\Delta P_j = P_{3j} - \frac{1}{d} \int_0^d P_{2j}(z) dz \qquad (3.60)$$

$\mathbf{P}_2(z)$ is the polarization in the transition layer and $\mathbf{P}_3$ is the polarization in the bulk of the transparent isotropic substrate. The y axis is chosen to be perpendicular to the plane of incidence. Since the polarization vector components $P_{2j}(z)$ and P_{3j} are proportional to the same electric field components of the incident wave, the parameters γ_j do not depend on the amplitude of the external exciting field. Being calculated in the zeroth order in d/λ, they do not depend on the wavelength of the light.[4] These quantities therefore characterize the optical properties of the transient layer to first order in d/λ. They are determined by the relative difference between the mean local field in the layer and the local field in the bulk medium. Therefore, Eqs (3.57) and (3.58) predict deviations from the two-phase Fresnel formulae even when there are no perturbations in the selvedge region, i.e., the surface is clean and the optical properties of atoms nearby the surface are identical with those in the bulk.

It should be noted that Eqs (3.57) and (3.58) have been derived without considering the electric field amplitude inside the layer as well as the relation between this field and the polarization vector. Consequently, they can also be applied when the optical response of the layer is nonlocal (see Section 3.2).

---

4) Strictly speaking, their dependence on λ is determined by the wavelength dependence of the index of refraction of the medium.

Equations (3.57) and (3.58) are more general as compared to Eqs (3.44) and (3.45). If the transition layer is macroscopic and its optical properties can be characterized by a dielectric function, they are reduced to the latter ones (see Problem 3.4).

As one can see from Eq. (3.60), the variation of the polarization vector $\mathbf{P}_2(z)$ within the layer is not important. Only the integral $\int_0^d \mathbf{P}_2(z)dz$ which represents the sum of atomic dipole moments per unit surface area is essential.

The explicit form of the γ_js can be found from a microscopic model of the transition layer in terms of the atomic polarizabilities α_{jj}. For example, in the case of a monolayer represented by a square lattice of atoms isotropic in the surface plane ($\alpha_{xx} = \alpha_{yy}$) on a simple cubic crystal with a lattice constant a, one finds (Sivukhin 1951)

$$\gamma_x = \gamma_y = \frac{a}{1 - (A/2a^3)\alpha_{xx}} \left[\frac{\alpha_0 - \alpha_{xx}}{\alpha_0} + \frac{\alpha_{xx}}{2a^3} \left(\frac{8\pi}{3} - A + B \right) \right] \quad (3.61)$$

and

$$\gamma_z = \frac{a}{1 + (A/a^3)\alpha_{zz}} \left[\frac{\alpha_0 - \alpha_{zz}}{\alpha_0} - \frac{\alpha_{zz}}{a^3} \left(\frac{8\pi}{3} - A + B \right) \right] \quad (3.62)$$

with $A = 9.035$, $B = 0.329 \cdot \exp(2\pi\Delta a/a)$ and α_0 the isotropic atomic polarizability of the substrate. The quantity $\Delta a = a - a_1$ takes into account a possible difference of the distance between the upper two atomic layers, a_1, from the lattice constant, and it is assumed that the selvedge region has the thickness a.

3.2
Nonlocal Optical Response

The derivation of the Fresnel equations has assumed that the electric displacement vector at any point in both bordering media is determined by the electric field amplitude at the same point, i.e., the dielectric response is local and is determined by the equation[5]

$$\mathbf{D}_\alpha(\mathbf{r}) = \epsilon_\alpha(\mathbf{r})\mathbf{E}_\alpha(\mathbf{r}) \quad (3.63)$$

where the subscript $\alpha = 1, 2$ refers to the different media and ϵ_α is the relevant dielectric function. In the close vicinity of the interface, the validity of relation (3.63) is questionable. Due to the continuity of the D_z component across the interface one has

$$\epsilon_1 E_{1z}(z = 0) = \epsilon_2 E_{2z}(z = 0) \quad (3.64)$$

which results in a discontinuity of E_z at $z = 0$ if E_z is not identically equal to zero.[6] As follows from Poisson's equation, this jump of the electric field is

---

5) Here and in the following we omit the argument ω.
6) Note that such a discontinuity does not occur for the electric field component parallel to the surface.

related to the induced surface charge density

$$\rho(\mathbf{r}) = \frac{1}{4\pi} \left(E_{2z} - E_{1z} \right) \delta(z) \tag{3.65}$$

whose spatial profile is described by the Dirac's δ-function. This result is obviously unphysical on a microscopic distance scale.

Hence there must be a problem in assuming a local relationship between the electric displacement and the electric field amplitude as given by Eq. (3.63). A rigorous treatment of the electromagnetic field within a few Ångstroms from the surface requires us to take into account that it varies essentially on the scale of the wavelength (Feibelman 1982). In this case, the optical response at the point $\mathbf{r}$ is related to the electric field at neighboring points $\mathbf{r}'$, i.e., it is *nonlocal*. Accordingly, instead of Eq. (3.63) one has to use the integral equation[7]

$$\mathbf{D}_\alpha(\mathbf{r}) = \int d\mathbf{r}' \hat{\epsilon}_\alpha(\mathbf{r},\mathbf{r}') : \mathbf{E}_\alpha(\mathbf{r}') \tag{3.66}$$

Note that in the case where the electric field varies slowly on the scale of the $\mathbf{r}'$-variation of $\hat{\epsilon}(\mathbf{r},\mathbf{r}')$, i.e., $\mathbf{E}_\alpha(\mathbf{r}') \approx \mathbf{E}_\alpha(\mathbf{r})$, Eq. (3.66) is reduced to the form of Eq. (3.63) with

$$\hat{\epsilon}_\alpha(\mathbf{r}) = \int d\mathbf{r}' \hat{\epsilon}_\alpha(\mathbf{r},\mathbf{r}') \tag{3.67}$$

To proceed further, we shall consider a simple system for which results can be obtained in a closed form. Let medium 1 be vacuum ($\epsilon_1 = 1$) and medium 2 be a jellium.[8] The jellium is invariant with respect to rotations around the surface normal and hence its dielectric tensor has the nonzero components $\epsilon_{xx} = \epsilon_{yy} \equiv \epsilon_\parallel$ and ϵ_{zz}.[9] Also, the translational invariance along the surface ensures that $\hat{\epsilon}(\mathbf{r},\mathbf{r}')$ depends on the difference between the lateral coordinates, i.e.,

$$\hat{\epsilon}(\mathbf{r},\mathbf{r}') = \hat{\epsilon}(\mathbf{r}_\parallel - \mathbf{r}'_\parallel, z, z') \tag{3.68}$$

In the absence of external charges the electric displacement vector satisfies the equation

$$\nabla \cdot \mathbf{D} = 0 \tag{3.69}$$

Substituting here the Fourier transform

$$\mathbf{D}(\mathbf{r}_\parallel, z) = \frac{1}{(2\pi)^2} \int d\mathbf{k}_\parallel \tilde{\mathbf{D}}(\mathbf{k}_\parallel, z) e^{i\mathbf{k}_\parallel \cdot \mathbf{r}_\parallel} \tag{3.70}$$

---

7) We take here into account that in general the dielectric response may be anisotropic. This implies that $\hat{\epsilon}$ is a tensor.

8) The so-called jellium model of a metal assumes that the positive charge of atomic cores is smeared out uniformly. One considers the electron density in the field of such a charge distribution.

9) We omit the subscript for medium 2 for brevity.

one derives the equation

$$ik_\| \cdot \tilde{\mathbf{D}}_\|(\mathbf{k}_\|, z) + \frac{d}{dz}\tilde{D}_z(\mathbf{k}_\|, z) = 0 \tag{3.71}$$

Due to the translational invariance along the surface, the quantity $\mathbf{k}_\|$ is conserved and Eq. (3.71) can be considered as a description of the optical response to the light illuminating the surface. In this case $\mathbf{k}_\|$ has the meaning of the wave vector component parallel to the surface. In the optical spectral range, $\mathbf{k}_\|$ is much less than the inverse distance on which the electron density varies at the surface. It is reasonable, therefore, to proceed to the long-wavelength limit, $\mathbf{k}_\| \to 0$, in Eq. (3.71). Then one concludes that the quantity $\tilde{D}_z(\mathbf{k}_\|, z)$ is constant across the interface and hence it is equal to the electric field amplitude in vacuum far from the surface, $\tilde{E}_z^{vac}(\mathbf{k}_\|)$.

On the other hand, the Fourier transform of Eq. (3.66) results in

$$\tilde{D}_j(\mathbf{k}_\|; z) = \int dz'\, \tilde{\epsilon}_{jj}(\mathbf{k}_\|; z, z')\tilde{E}_j(\mathbf{k}_\|; z'), \quad j = x, y, z \tag{3.72}$$

where $\tilde{E}_j$ is the Fourier transform of E_j and

$$\tilde{\epsilon}_{jj}(\mathbf{k}_\|; z, z') = \int d\mathbf{x}\, \epsilon_{jj}(\mathbf{x}, z, z')e^{i\mathbf{k}_\| \cdot \mathbf{x}} \tag{3.73}$$

Finally, one obtains the equation

$$\int dz'\, \tilde{\epsilon}_{zz}(\mathbf{k}_\|; z, z')\tilde{E}_z(\mathbf{k}_\|; z') = \tilde{E}_z^{vac}(\mathbf{k}_\|) \tag{3.74}$$

which completely determines $\tilde{E}_z(\mathbf{k}_\|; z')$ at small $\mathbf{k}_\|$. It generalizes the matching condition $D_{1z}(z = 0) = D_{2z}(z = 0)$ to the case of a nonlocal optical response and provides a continuous solution for the electric field component normal to the surface.

The account of the nonlocal character of the dielectric response modifies the Fresnel equations. In the framework of the jellium model for a metal, the reflection coefficients in s- and p-polarizations can be represented as[10]

$$r_s = r_s^{loc}\left[1 + 2ik_{iz}d_\|(\omega)\right] \tag{3.75}$$

and

$$r_p = r_p^{loc}\left\{1 - \frac{2ik_{iz}(\epsilon_b - 1)}{k_{tz}^2 - \epsilon_b^2 k_{iz}^2}\left[k_{tz}^2 d_\|(\omega) - \epsilon_b k_{i\|}^2 d_\perp(\omega)\right]\right\} \tag{3.76}$$

respectively. Here r_s^{loc} and r_p^{loc} are the Fresnel reflection coefficients, Eqs (3.16) and (3.22), in the local approximation, $k_{i\|}$ and k_{iz} are the components of the incident wave vector, Eqs (3.5) and (3.6), k_{tz} is the z-component of the refracted wave vector, Eq. (3.7), and ϵ_b is the bulk dielectric function given by Eq. (2.64).

[10] One can derive these formulae by solving Maxwell's equations along with Eq. (3.74) in the long-wavelength limit and keeping the terms up to the first order in $\mathbf{k}_\|$.

The quantities $d_\parallel(\omega)$ and $d_\perp(\omega)$ are found to be

$$d_\parallel(\omega) = \left[\int_{-\infty}^{\infty} dz \frac{d\bar{\epsilon}_\parallel(z)}{dz}\right]^{-1} \int_{-\infty}^{\infty} dz\, z \frac{d\bar{\epsilon}_\parallel(z)}{dz} \tag{3.77}$$

and[11]

$$d_\perp(\omega) = \left[\int_{-\infty}^{\infty} dz \rho(z)\right]^{-1} \int_{-\infty}^{\infty} dz\, z\rho(z) \tag{3.78}$$

where $\rho(z)$ is the density of the surface charge induced by the external electromagnetic field. The function $\bar{\epsilon}_\parallel(z)$[12] interpolates smoothly between 1 in vacuum and ϵ_b in the bulk jellium and its derivative has a peak near the surface. The quantity $\mathrm{Re}[d_\parallel(\omega)]$, therefore, measures the surface position. Analogously, $\mathrm{Re}[d_\perp(\omega)]$ measures the centroid of the induced surface charge. In the case where there is no bulk power absorption, i.e., ϵ_b is real, the corresponding imaginary parts, $\mathrm{Im}[d_\parallel(\omega)]$ and $\mathrm{Im}[d_\perp(\omega)]$, determine the power absorption at the surface associated with the tangential and normal components of the electric field, respectively. Note that if the interface could be described by a three-phase model where the region of the $\bar{\epsilon}_\parallel$ variation plays the role of an isotropic transition layer with effective dielectric function and effective thickness d, the results given by Eqs (3.75) and (3.76) are reduced to Eqs (3.44) and (3.45), respectively (see Problem 3.5).

It is also worthwhile to note that Eqs (3.75) and (3.76) derived for a jellium which is isotropic in the surface plane are identical with Eqs (3.57) and (3.58) if one assumes that the transition layer is isotropic in the xy plane, i.e., $\gamma_x = \gamma_y$. Comparing the two sets of equations and taking into account the definitions of the quantities k_{iz}, $k_{i\parallel}$ and k_{tz}, reveals that the quantities $d_\parallel$ and $d_\perp$ coincide with the parameters $\gamma_x = \gamma_y$ and γ_z, respectively. This conclusion is not surprising since the microscopic theory considered in Section 3.1.3 has defined the transition layer as a layer where the polarization differs from that in the bulk medium, i.e., it indirectly includes the effect of nonlocality in the selvedge region.

---

11) This expression for $d_\perp(\omega)$ is valid for frequencies below the plasma frequency of the metal.

12) The function $\bar{\epsilon}_\parallel(z)$ is determined as follows

$$\bar{\epsilon}_\parallel(z) = \int_{-\infty}^{\infty} dz'\, \bar{\epsilon}_\parallel(\mathbf{k}_\parallel = 0, z, z') \tag{3.79}$$

Also, it is implied that a few Ångstroms apart from the jellium the quantities $\epsilon_{jj}(\mathbf{r}_\parallel - \mathbf{r}'_\parallel, z, z')$ can be extrapolated by the dielectric tensor components in vacuum, $\delta(\mathbf{r}_\parallel - \mathbf{r}'_\parallel)\delta(z - z')$.

3.3
Surface Polaritons

3.3.1
Fundamental Properties

In Section 3.1 we have considered reflected and transmitted waves which appear when a plane electromagnetic wave hits the boundary between two media. We have seen that propagation of the wave in medium 2 obeys Eq. (3.4). In the frequency range corresponding to elementary substrate excitations where the dispersion of ϵ_2 is remarkable, one has to take into account its frequency dependence, i.e.,

$$k_t = \frac{\omega}{c}\sqrt{\epsilon_2(\omega)} \tag{3.80}$$

This equation represents the dispersion law of coupled modes arising from interactions between light and substrate excitations which are called *polaritons*. As follows from relation (3.80), these polaritons can propagate in the medium at the frequency ω if $\epsilon_2(\omega) > 0$.

However, in some frequency ranges, the quantity $\epsilon_2(\omega)$ may be negative. This occurs, for example, at frequencies below the plasma frequency in metals or n-type semiconductors, and between the frequencies of transverse and longitudinal optical phonons or excitons in dielectrics and semiconductors. In this case, k_t becomes pure imaginary, which means that polaritons cannot exist in medium 2. Nevertheless a similar coupled mode can propagate *along* the crystal surface, not penetrating deeply into the substrate. The corresponding solutions of Maxwell's equations are called *surface polaritons* (SPs).

Any electromagnetic wave travelling along the interface can be represented as a superposition of two independently polarized components, namely a transverse magnetic (TM) wave and a transverse electric (TE) wave. Let us choose the x-axis along the wave vector $\mathbf{q}$. Then in a TM wave the electric and magnetic field vectors have components $(E_x, 0, E_z)$ and $(0, H_y, 0)$, respectively. A TE wave is represented by the components $(0, E_y, 0)$ and $(H_x, 0, H_z)$. We shall seek the solution of Maxwell's equations corresponding to SPs in the form

$$\mathbf{E}_1(\mathbf{r}, t) = \mathbf{E}_{10}e^{iqx}e^{\kappa_1 z}e^{-i\omega t}, \qquad z < 0 \tag{3.81}$$

$$\mathbf{E}_2(\mathbf{r}, t) = \mathbf{E}_{20}e^{iqx}e^{-\kappa_2 z}e^{-i\omega t}, \qquad z > 0 \tag{3.82}$$

where both κ_1 and κ_2 are real and positive to ensure finiteness of the solution at infinite distance from the interface. The magnetic field vector can be written in a similar form on both sides of the boundary.

Let us consider first the case of TM polarization. The substitution of the expressions (3.81) and (3.82) into the equations for the electric field vector[13]

$$\nabla \times \nabla \times \mathbf{E} = -\frac{\epsilon_j}{c^2}\frac{\partial^2 \mathbf{E}}{\partial t^2}, \quad j = 1, 2 \tag{3.83}$$

gives the relations

$$\kappa_1^2 = q^2 - \epsilon_1 \frac{\omega^2}{c^2} \tag{3.84}$$

$$\kappa_2^2 = q^2 - \epsilon_2 \frac{\omega^2}{c^2} \tag{3.85}$$

Further, from the equation $\nabla \times \mathbf{H} = (1/c)(\partial \mathbf{D}/\partial t)$ one finds

$$\kappa_1 H_{10y} = -i\frac{\omega}{c}\epsilon_1 E_{10x} \tag{3.86}$$

$$\kappa_2 H_{20y} = i\frac{\omega}{c}\epsilon_2 E_{20x} \tag{3.87}$$

From here, taking into account the continuity of the tangential components across the interface, we derive the equation

$$\frac{\kappa_1}{\kappa_2} = -\frac{\epsilon_1}{\epsilon_2} \tag{3.88}$$

Now, combining Eq. (3.88) with (3.84) and (3.85), we obtain the dispersion law for SPs as

$$q = \frac{\omega}{c}\sqrt{\frac{\epsilon_1 \epsilon_2}{\epsilon_1 + \epsilon_2}} \tag{3.89}$$

The consideration of TE waves shows that SPs cannot exist with such a polarization (see Problem 3.6).

As follows from the dispersion relation (3.89), that the frequency range where SPs can exist is narrower than that given by the condition $\epsilon_2(\omega) < 0$: it is also necessary to fulfil the inequality[14]

$$\epsilon_2(\omega) < -\epsilon_1 \tag{3.90}$$

When $q \to \infty$ the dispersion curve $\omega(q)$ asymptotically approaches the frequency ω_s which satisfies the equation $\epsilon_2(\omega_s) = -\epsilon_1$ and corresponds to elementary surface excitations in the nonretarded limit (see Sections 2.1.3 and 2.1.4). The realness of κ_1 and also Eq. (3.84) dictate that the modulus of the SP wave vector, q, is always greater than $(\omega/c)\sqrt{\epsilon_1}$ whereas in light waves

13) This equation follows from Maxwell's equations
   $\nabla \times \mathbf{E} = -(1/c)(\partial \mathbf{H}/\partial t)$ and $\nabla \times \mathbf{H} = (1/c)(\partial \mathbf{D}/\partial t)$.
14) We assume for simplicity that in the considered frequency range
   $\epsilon_1 > 0$.

travelling in medium 1 the wave vector component parallel to the surface, k_x, is always less than this quantity. In other words, q and k_x do not match each other at all incidence angles of light. This means that SPs cannot be excited by light or be transformed into light. The SP represents a nonradiative mode. However, this conclusion is valid only for a flat interface between two semi-infinite media. Some special experimental methods allowing us to generate SPs will be considered below (Section 3.3.4).

The quantities κ_1 and κ_2 determine the *penetration depths*, δ_1 and δ_2, of SPs into media 1 and 2, respectively. Using Eqs (3.84), (3.85) and (3.89) they can be written as

$$\delta_1 = \frac{1}{\kappa_1} = \frac{\lambda}{2\pi}\sqrt{\frac{\epsilon_1 + \epsilon_2}{-\epsilon_1^2}} \tag{3.91}$$

$$\delta_2 = \frac{1}{\kappa_2} = \frac{\lambda}{2\pi}\sqrt{\frac{\epsilon_1 + \epsilon_2}{-\epsilon_2^2}} \tag{3.92}$$

Their ratio is given by

$$\frac{\delta_1}{\delta_2} = \frac{|\epsilon_2|}{\epsilon_1} \tag{3.93}$$

Relaxation processes in the bordering media introduce imaginary parts of ϵ_1 and ϵ_2. As a result, the SP wave vector (3.89) becomes complex as well. Then one can define the *propagation length*, L, after which the SP intensity decreases to $1/e$:

$$L = (2\,\mathrm{Im}\,q)^{-1} \tag{3.94}$$

3.3.2
Surface Plasmon Polaritons

Depending on the substrate excitations which are coupled with light into the SP mode, one distinguishes surface plasmon polaritons, surface phonon polaritons, surface exciton polaritons, etc. In this section we shall consider surface plasmon polaritons in some detail. This type of electromagnetic wave was first discussed by Sommerfeld in connection with the propagation of radiowaves along the Earth's surface (Sommerfeld 1909).

The optical response of conduction electrons in metals or n-type semiconductors can be described by Drude's dielectric function, Eq. (2.64). Then from Eq. (3.90) we obtain that SP can exist in the frequency range

$$\omega < \frac{\omega_p}{\sqrt{1 + \epsilon_1}} \tag{3.95}$$

In metals, the plasmon energy, $\hbar\omega_p$, is of the order of 10 eV and hence surface plasmon polaritons can be excited up to the ultraviolet spectral region. In semiconductors, the plasma frequency is determined by the doping level.

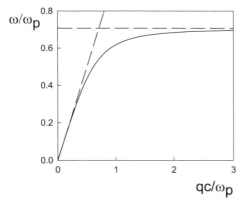

Fig. 3.3 The dispersion curve of a surface plasmon polariton at a metal–vacuum interface. The inclined straight line corresponds to the dispersion of light propagating in vacuum, parallel to the metal surface. The horizontal line represents the frequency of the surface plasmon polariton in the nonretarded limit, $\omega_p / \sqrt{2}$.

For a conduction electron density of 10^{17} cm^{-3}, the corresponding plasmon energies fall in the range 10–30 meV.

Figure 3.3 shows the dispersion law of surface plasmon polaritons at a metal–vacuum interface ($\epsilon_1 = 1$). The curve $\omega(q)$ is located to the right from the light line indicating that a SP cannot be excited by light. Its limiting value $\omega_s = \omega_p / \sqrt{2}$ corresponds to the frequency of surface plasmons (see Section 2.1.3).

At frequencies well below the plasma frequency, the absolute value of ϵ_2 sufficiently exceeds unity. In this case, as follows from Eq. (3.93), the SP field penetrates into metal over much shorter distances than into vacuum. For example, at $\lambda = 6000$ Å for silver, $\delta_1 = 3900$ Å and $\delta_2 = 240$ Å, and for gold 2800 Å and 310 Å, respectively. The SP propagation length depends to a great extent on its frequency. In silver, L is equal to 22 μ at $\lambda = 5145$ Å and is much larger, 0.05 cm, at $\lambda = 10600$ Å (Raether 1988).

3.3.3
Surface Phonon Polaritons

Let us turn now to the case when light is coupled with optical phonons in an ionic crystal. The corresponding dielectric function can be written in two equivalent forms

$$\epsilon_2(\omega) = \epsilon_\infty + (\epsilon_0 - \epsilon_\infty) \frac{\omega_{TO}^2}{\omega_{TO}^2 - \omega^2} = \frac{\omega_{LO}^2 - \omega^2}{\omega_{TO}^2 - \omega^2} \epsilon_\infty \qquad (3.96)$$

where ϵ_0 and ϵ_∞ are the static and high-frequency dielectric functions, respectively, ω_{TO} is the frequency of transverse optical phonons and ω_{LO} is the fre-

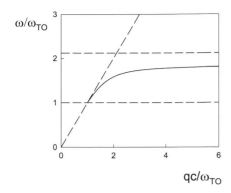

ω/ω_{TO}

qc/ω_{TO}

Fig. 3.4 Dispersion curve of a surface phonon polariton at a crystal–vacuum interface calculated for $\epsilon_0 = 9$ and $\epsilon_\infty = 2$. The inclined straight line corresponds to the dispersion of light propagating in vacuum parallel to the crystal surface. The two horizontal lines show the frequencies ω_{TO} and ω_{LO}.

quency of longitudinal optical phonons. The latter quantity determines the null of $\epsilon_2(\omega)$ and satisfies the relation

$$\omega_{LO} = \omega_{TO}\sqrt{\frac{\epsilon_0}{\epsilon_\infty}} \qquad (3.97)$$

The condition (3.90) for the existence of SPs is then fulfilled in the frequency range

$$\omega_{TO} < \omega < \omega_s \qquad (3.98)$$

where the limiting value

$$\omega_s = \omega_{TO}\sqrt{\frac{\epsilon_0 + \epsilon_1}{\epsilon_\infty + \epsilon_1}} \qquad (3.99)$$

corresponds to the frequency of a surface optical phonon in the nonretarded limit (see Section 2.1.4).

Figure 3.4 shows the dispersion curve of a surface phonon polariton. If the considered frequency range contains several optical phonon modes, Eq. (3.96) must be generalized to a sum over them. Another complication arises when the plasma frequency in a doped semiconductor crystal has the same order of magnitude as the optical phonon frequencies. Then one has to add into Eq. (3.96) the contribution of conduction electrons given by $-\omega_p^2/\omega^2$.

3.3.4
Excitation of Surface Polaritons

We have seen that SPs are nonradiative modes and that they cannot be excited by incident light at a flat interface between two media. This follows from

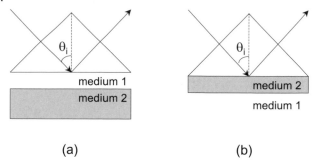

(a)   (b)

Fig. 3.5 Geometry for obtaining surface polariton excitations. (a) Otto configuration. A prism of refractive index n is suspended above medium 2. (b) Kretschmann configuration. The sample film is deposited onto the prism surface.

the fact that the SP wave vector, q, is longer than the wave vector of light in medium 1, $k = (\omega/c)\sqrt{\epsilon_1}$. However, this restriction can be removed at more complex interfaces.

Otto configuration

One can "elongate" the wave vector of light by passing it through a medium (a prism) with a refractive index $n > \sqrt{\epsilon_1}$ (*Otto configuration* (Otto 1968)) (see Fig. 3.5a). In this case, the wave vector component k_x does not change across both interfaces, prism/medium 1 and medium 1/medium 2, and is given by

$$k_x = \frac{\omega}{c} n \sin \theta_i \tag{3.100}$$

where θ_i is the incidence angle of light at the interface between the prism and medium 1. On the other hand, to match this quantity with the SP wave vector one must have

$$k_x = q > \frac{\omega}{c}\sqrt{\epsilon_1} \tag{3.101}$$

In medium 1, the wave vector component perpendicular to the interface is found as

$$k_{1z} = \sqrt{\frac{\omega^2}{c^2}\epsilon_1 - k_x^2} \tag{3.102}$$

Now, combining Eqs (3.101) and (3.102), we conclude that k_{1z} is purely imaginary, i.e., the wave in medium 1 is evanescent. In other words, to excite SP it is necessary to realize total internal reflection at the prism/medium 1 interface, i.e.,

$$\theta_i > \arcsin\left(\frac{\sqrt{\epsilon_1}}{n}\right) \tag{3.103}$$

The thickness of the gap between the prism and the sample (medium 2) should be of the order of the wavelength to ensure the delivery of the electromagnetic field to the sample surface.

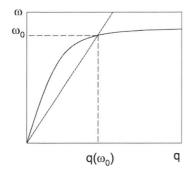

$q(\omega_0)$ q

Fig. 3.6 Scheme illustrating surface polariton excitation in total internal reflection configurations. The curved line represents the dispersion of a surface polariton. The inclined straight line corresponds to the dispersion of light propagating in the prism and is described by the relation $\omega = ck_x/(n \sin \theta_i)$. Its intersection with the dispersion curve determines the incidence angle θ_{i0} at which a surface polariton will be excited.

The condition for SP excitation, Eq. (3.101), can be rewritten as

$$n \sin \theta_i = \sqrt{\frac{\epsilon_1 \epsilon_2(\omega)}{\epsilon_1 + \epsilon_2(\omega)}} \qquad (3.104)$$

If the light frequency is fixed at ω, the angle θ_i at which a SP can be excited, can be determined from this relation. On the other hand, if one tunes the light frequency keeping the incidence angle fixed at θ_i, this equation gives the frequency ω corresponding to the excitation of a SP. This is illustrated in Fig. 3.6 with the help of the dispersion lines.

Kretschmann configuration

The Otto configuration is well adapted for the cases when it is necessary to avoid contact between prism and surface. However, similar conditions can be realized if the sample film (medium 2) is deposited directly onto the prism surface. Such a geometry is known as a *Kretschmann configuration* (Kretschmann and Raether 1968) (Fig. 3.5b). The film thickness in this case must be less than the penetration depth of the evanescent wave excited at the prism/film interface which is usually of the order of a few 100 Å.

Excitation on a grating

Light incident on a grating perpendicularly to its ruling at an angle θ_i will be diffracted into diffraction maxima at angles θ_m determined by the equation (Born and Wolf 1975)

$$d(\sin \theta_i - \sin \theta_m) = \pm m\lambda \qquad (3.105)$$

with d the grating period and m an integer. This relation can be rewritten as

$$k_{mx} = k_{ix} \pm \frac{2\pi m}{d} \qquad (3.106)$$

where k_{ix} and k_{mx} are the wave vector components along the grating of the incident wave and of the diffracted wave corresponding to the mth maximum, respectively. This means that, upon diffraction, the tangential wave vector component acquires shifts given by $\pm 2\pi m/d$. Therefore, the matching condition between light and SP,

$$\frac{\omega}{c}\sqrt{\epsilon_1}\sin\theta_i \pm \frac{2\pi m}{d} = q \tag{3.107}$$

can be satisfied for some θ_i and m. This transformation is illustrated in Fig. 3.7.

A similar mechanism of SP excitation occurs when light strikes a rough surface (see Section 3.4). The representation of the surface relief by the Fourier integral,[15]

$$\zeta(\mathbf{r}_\parallel) = \frac{1}{(2\pi)^2}\int \hat{\zeta}(\mathbf{G})e^{i\mathbf{G}\cdot\mathbf{r}_\parallel}d\mathbf{G}$$
$$= \frac{1}{(2\pi)^2}\int \left\{\mathrm{Re}[\hat{\zeta}(\mathbf{G})]\cos(\mathbf{G}\cdot\mathbf{r}_\parallel) - \mathrm{Im}[\hat{\zeta}(\mathbf{G})]\sin(\mathbf{G}\cdot\mathbf{r}_\parallel)\right\}d\mathbf{G} \tag{3.108}$$

can be considered as a superposition of an infinite set of sinusoidal gratings with the periods along the x axis given by $d(\mathbf{G}) = 2\pi/G_x$. Then the matching condition for excitation of SPs propagating in the plane of incidence can be fulfilled for a grating specified by the vector $\mathbf{G}$ which satisfies the relation[16]

$$\frac{\omega}{c}\sqrt{\epsilon_1}\sin\theta_i \pm G_x = q \tag{3.109}$$

Reflectivity at SP excitation
A SP carries electromagnetic energy and therefore its excitation can be detected by measuring the intensity of the reflected light. We consider for definiteness the reflectivity in the Kretschmann configuration. It can be described by Fresnel equations for a three-layer system (see Section 3.1.3). Let E_{i0} be the electric field amplitude of the p-polarized wave incident onto the prism/film interface and E_{r0} be that of the wave reflected from this interface. Then

$$R = \left|\frac{E_{r0}}{E_{i0}}\right|^2 = \left|\frac{r_{02} + r_{21}\exp(2ik_{2z}d)}{1 + r_{02}r_{21}\exp(2ik_{2z}d)}\right|^2 \tag{3.110}$$

where r_{02} and r_{21} are the reflection coefficients for p-polarized light at the interfaces between the prism and the film, and between the film and medium 1, respectively, k_{2z} is the z-component of the wave vector in the film and d is

---

15) We have taken into account here the realness of the function $\zeta(\mathbf{r}_\parallel)$.
16) The same relation describes the reverse process, namely the transformation of a SP into scattered light.

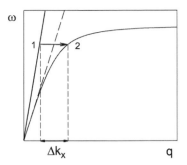

Fig. 3.7 Excitation of surface polaritons on a grating. The solid inclined line represents the dispersion of the incident light, $\omega = ck/\sin\theta_i$, whereas the dashed inclined line is the boundary light line corresponding to $\theta_i = \pi/2$. The matching between incident light and surface polariton dispersions occurs due to the transformation of the wave vector of light on diffraction. This process corresponds to a transition from point 1 to point 2 and is accompanied by an increase in the wave vector of light by $\Delta k_x = 2\pi m/d$.

the film thickness. Under special conditions, when $|\operatorname{Re}(\epsilon_2)| \gg 1, |\operatorname{Im}(\epsilon_2)|$, the reflectivity as a function of k_x can be approximated near the resonance $k_x = q$ by the Lorentzian contour (Raether 1988)

$$R = 1 - \frac{4\Gamma_i\Gamma_{\text{rad}}}{[k_x - (q + \Delta k_x)]^2 + (\Gamma_i + \Gamma_{\text{rad}})^2} \tag{3.111}$$

where q is the SP wave vector given by Eq. (3.89) and $\Gamma_i = \operatorname{Im}(q)$ is the so-called internal damping. The quantities Δk_x and Γ_{rad} determine the shift of the resonance $k_x = q$ and its broadening, respectively, due to the presence of the prism. At incidence angles far from that given by the matching condition (3.104) the reflectivity approaches unity due to total internal reflection. The reflectivity decrease near the resonance is therefore called *attenuated total reflection* (ATR). This behavior is shown in Fig. 3.8. The reflectivity has a zero minimum value if $\Gamma_{\text{rad}} = \Gamma_i$. This condition fixes the thickness of the sample film at a given wavelength. For example, at $\lambda = 5000$ Å ATR on a silver film gives $R = 0$ at $d = 550$ Å (Raether 1988).

3.3.5
Electromagnetic Field Enhancement

Energy conservation suggests that when the reflectivity at SP excitation is minimal, the intensity of the SP electromagnetic field has its maximum. For the Kretschmann configuration, the intensity enhancement, η, which is obtained in SP is given by the ratio of the SP intensity in medium 1 to the intensity of light incident onto the prism/film interface. It is found as (Raether 1988)

$$\eta = \left|\frac{E_{sp,1}}{E_{i0}}\right|^2 = \frac{n^2}{\epsilon_1}\left|\frac{t_{02}t_{21}\exp(ik_{2z}d)}{1 + r_{02}r_{21}\exp(2ik_{2z}d)}\right|^2 \tag{3.112}$$

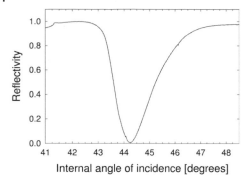

Fig. 3.8 Reflectivity in the vicinity of the resonance corresponding to a SP excitation. The experimental data are for a 45 nm thick Au film deposited onto a quartz prism. The increase in reflectivity at 41.3° corresponds to the Brewster angle.

where t_{02} and t_{21} are the ratios of the magnetic field amplitude in the transmitted wave to that in the incident p-polarized wave at the prism/film and film/medium 2 interfaces, respectively,[17] and the other notations are the same as in Eq. (3.110). If $|\operatorname{Re}(\epsilon_2)| \gg 1, |\operatorname{Im}(\epsilon_2)|$, then in the vicinity of the resonance $k_x = q$ (corresponding to SP excitation) Eq. (3.112) can be represented in the form of a Lorentzian contour

$$\eta = \eta_{\max} \frac{4\Gamma_i \Gamma_{\mathrm{rad}}}{[k_x - (q + \Delta k_x)]^2 + (\Gamma_i + \Gamma_{\mathrm{rad}})^2} \qquad (3.113)$$

where $\eta_{\max}$ is the maximum value of enhancement which occurs at the exact resonance $k_x = q + \Delta k_x$ when $\Gamma_{\mathrm{rad}} = \Gamma_i$.

Electromagnetic field enhancement varies with the wavelength and can be rather strong. For example, if $\epsilon_1 = 1$ and $n = \sqrt{2.2}$ (quartz) one finds for silver at $\lambda = 4500$ Å $\eta_{\max} \approx 100$; at $\lambda = 6000$ Å $\eta_{\max} \approx 200$; and at $\lambda = 7000$ Å $\eta_{\max} \approx 250$. For other substances the enhancement is somewhat lower: at $\lambda = 6000$ Å $\eta_{\max}$ is about 30 for gold, about 40 for aluminium and about 7 for copper (Raether 1988).

It should be noted, however, that if one calculates the enhancement factor relative to the intensity of the light before it enters the prism, i.e., in the surrounding medium 1, η has to be exchanged by $(\epsilon_1/n^2)\eta$ and thus the enhancement is somewhat smaller.

3.3.6
Localized Surface Polaritons

Surface polaritons can also exist on curved and closed surfaces, e.g., on spheres and ellipsoids. In the latter case, SPs are called localized and the

17) The quantities t_{ij} can be found from the Fresnel equations (see Section 3.1.1).

SP dispersion relation depends to a great extent on the surface geometry (see, e.g., (Raether 1988)).

In the nonretarded limit ($c \to \infty$) the SP modes of a sphere having the dielectric function $\epsilon_2(\omega)$ are given by the relation

$$\epsilon_2(\omega) = -\epsilon_1 \frac{l+1}{l}, \quad l = 1, 2, 3, \ldots \tag{3.114}$$

where ϵ_1 is the dielectric constant of the surrounding medium. The modes denoted by l are radiative, i.e., they can be excited by light or transferred into light without any additional devices.

The frequency of the lowest mode corresponding to $l = 1$ obeys

$$\epsilon_2(\omega) = -2\epsilon_1 \tag{3.115}$$

which can also be deduced from solving the standard electrostatic problem. For this, let us assume that the sphere radius is much less than the wavelength of the incident light. Then the field inside the sphere is uniform and is given by (Stratton 1941)

$$\mathbf{E}_{int} = \frac{3\epsilon_1}{\epsilon_2(\omega) + 2\epsilon_1} \mathbf{E}_{i0} \tag{3.116}$$

where $\mathbf{E}_{i0}$ is the electric field amplitude of the incident light. The zero of the denominator in Eq. (3.116) indicates resonant enhancement of the field which takes place at the same frequency as that satisfying Eq. (3.115).

The field at the external sphere surface can be found from the boundary conditions (see Fig. 3.9)

$$E_{ex,\parallel} = E_{int,\parallel} = \frac{3\epsilon_1}{\epsilon_2(\omega) + 2\epsilon_1} E_{i0} \sin \theta \tag{3.117}$$

$$\epsilon_1 E_{ex,\perp} = \epsilon_2(\omega) E_{int,\perp} = \epsilon_2(\omega) \frac{3\epsilon_1}{\epsilon_2(\omega) + 2\epsilon_1} E_{i0} \cos \theta \tag{3.118}$$

where the subscripts $\parallel$ and $\perp$ denote the field components tangential and normal to the sphere surface, respectively, and θ is the angle which the electric field inside the sphere has with respect to its radius.

If one takes into account that the dielectric function of the sphere is a complex quantity, i.e., $\epsilon_2 = \epsilon_2' + i\epsilon_2''$, and $|\epsilon_2''| \ll |\epsilon_2'|$, Eq. (3.118) gives a maximum enhancement factor for $\theta = 0$ at the resonance $\epsilon_2'(\omega) = -2\epsilon_1$:

$$\eta_{max} = \left| \frac{E_{ex}}{E_{i0}} \right|^2 = \left(\frac{3\epsilon_2'}{\epsilon_2''} \right)^2 \tag{3.119}$$

For a silver sphere embedded in a vacuum ($\epsilon_1 = 1$) the resonance condition (3.115) gives $\lambda = 3500$ Å with a corresponding $\epsilon_2'' = 0.28$. This results in

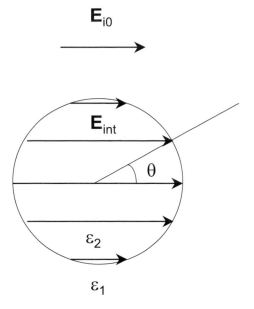

Fig. 3.9 Electric fields inside and outside a small sphere. See text.

$\eta_{max} \approx 480$, somewhat higher than for a flat silver/vacuum interface at the same wavelength.

In the case of a spheroid surrounded by a vacuum the electric field at the tip of its major axis oriented parallel to the electric field of light is given by

$$E_{tip} = \frac{\epsilon_2(\omega)}{1 + [\epsilon_2(\omega) - 1]A} E_{i0} \tag{3.120}$$

where A is the depolarization factor and it is assumed that the wavelength is much larger than both spheroid axes, a and b. Now the resonance condition has the form

$$\epsilon_2'(\omega) = 1 - \frac{1}{A} \tag{3.121}$$

For a sphere $A = 1/3$, this equation coincides with Eq. (3.115) with $\epsilon_1 = 1$. The enhancement factor at the resonance is found as

$$\eta_{max} = \left| \frac{E_{tip}}{E_{i0}} \right|^2 = \left(\frac{\epsilon_2'}{\epsilon_2'' A} \right)^2 \tag{3.122}$$

From here one concludes that the smaller A (the more elongated the ellipsoid), the greater the enhancement factor. This field enhancement is similar to the well-known "lightning-rod effect". For example, when $A = 0.1$ ($b/a \approx 1/3$) the condition (3.121) gives $\epsilon_2' = -9$. For silver this equality holds at $\lambda \approx 4900$ Å and correspondingly $\epsilon_2'' = 0.3$, which results in $\eta_{max} = 9 \cdot 10^4$.

3.4
Scattering of Light at Rough Surfaces

In this section we shall consider how the results obtained above for reflection from a plane surface are modified in the case of rough surfaces. We shall assume that the amplitude of surface roughness, $\zeta(\mathbf{r}_{\parallel})$, can be treated as small[18] and thus the solutions of Maxwell's equations can be expanded as a Taylor series in it (Maradudin and Mills 1975). We suppose for simplicity that above the surface $z = \zeta(\mathbf{r}_{\parallel})$ is vacuum, while below it is the isotropic medium with a complex frequency-dependent dielectric function $\epsilon = \epsilon(\omega)$. The total dielectric function can then be written as

$$\epsilon(z) = \Theta(z - \zeta(\mathbf{r}_{\parallel})) + \epsilon\Theta(\zeta(\mathbf{r}_{\parallel}) - z) \tag{3.123}$$

where

$$\Theta(z) = \begin{cases} 1, & z > 0 \\ 0, & z < 0 \end{cases} \tag{3.124}$$

is Heaviside's unit step function. Expanding the function $\epsilon(z)$ in powers of $\zeta(\mathbf{r}_{\parallel})$, we get

$$\epsilon(z) = \epsilon_0(z) + (\epsilon - 1)\zeta(\mathbf{r}_{\parallel})\delta(z) + O(\zeta^2) \tag{3.125}$$

where

$$\epsilon_0(z) = \begin{cases} 1, & z > 0 \\ \epsilon, & z < 0 \end{cases} \tag{3.126}$$

and we have used the relation

$$\frac{d\Theta(z)}{dz} = \delta(z) \tag{3.127}$$

The electric field $\mathbf{E}(\mathbf{r}, t) = \mathbf{E}(\mathbf{r})\exp(-i\omega t)$ satisfies the equation[19]

$$\nabla^2 \mathbf{E}(\mathbf{r}) - \nabla[\nabla \cdot \mathbf{E}(\mathbf{r})] + \epsilon(z)\frac{\omega^2}{c^2}\mathbf{E}(\mathbf{r}) = 0 \tag{3.128}$$

Its solution can be expanded in a power series in ζ

$$\mathbf{E} = \mathbf{E}^{(0)} + \mathbf{E}^{(1)} + \dots \tag{3.129}$$

Each term in this expansion obeys the equations which are obtained by substituting (3.125) and (3.129) into Eq. (3.128) and collecting the terms of the same order. It is easy to see that the zeroth-order equation[20]

$$\nabla^2 \mathbf{E}^{(0)}(\mathbf{r}) + \epsilon_0(z)\frac{\omega^2}{c^2}\mathbf{E}^{(0)}(\mathbf{r}) = 0 \tag{3.130}$$

18) This implies that the corresponding root-mean-square height of the surface, σ, is much less than the wavelength of light.
19) This equation follows from Maxwell's equations; see (Born and Wolf 1975) for details.
20) We exclude here the plane $z = 0$ and take into account that $\nabla \cdot \mathbf{E} = [1/\epsilon_0(z)]\nabla \cdot \mathbf{D} = 0$.

describes the electromagnetic field at a flat boundary. Its solutions are given by the Fresnel formulae (see Section 3.1.1), i.e., are represented by the electromagnetic wave reflected in the specular direction and by the refracted wave in the medium. The terms of higher orders in Eq. (3.129) give the electromagnetic field scattered at surface roughness.

The first-order equation has the form

$$\nabla^2 \mathbf{E}^{(1)}(\mathbf{r}) - \nabla[\nabla \cdot \mathbf{E}^{(1)}(\mathbf{r})] + \epsilon_0(z)\frac{\omega^2}{c^2}\mathbf{E}^{(1)}(\mathbf{r})$$

$$= -(\epsilon - 1)\zeta(\mathbf{r}_\parallel)\delta(z)\frac{\omega^2}{c^2}\mathbf{E}^{(0)}(\mathbf{r}) \tag{3.131}$$

Let us introduce now the tensorial Green's function $\hat{\mathbf{G}}(\mathbf{r}, \mathbf{r}')$ which satisfies the matrix equation

$$\nabla^2 \hat{\mathbf{G}}(\mathbf{r}, \mathbf{r}') - \nabla[\nabla \cdot \hat{\mathbf{G}}(\mathbf{r}, \mathbf{r}')] + \epsilon_0(z)\frac{\omega^2}{c^2}\hat{\mathbf{G}}(\mathbf{r}, \mathbf{r}') = 4\pi\delta(\mathbf{r} - \mathbf{r}')\hat{I} \tag{3.132}$$

where $\hat{I}$ is the unit 3×3 matrix. Then the first-order correction can be written as

$$\mathbf{E}^{(1)}(\mathbf{r}) = -(\epsilon - 1)\frac{\omega^2}{4\pi c^2}\int d\mathbf{r}'\zeta(\mathbf{r}'_\parallel)\delta(z')\hat{\mathbf{G}}(\mathbf{r}, \mathbf{r}') : \mathbf{E}^{(0)}(\mathbf{r}') \tag{3.133}$$

Equation (3.133) gives the electric field scattered by surface roughness in the linear approximation with respect to the roughness profile. The conditions which one imposes on the Green's function at the plane $z = 0$ ensure that $\mathbf{E}^{(1)}(\mathbf{r})$ also obeys the necessary boundary conditions.

For the further consideration it is convenient to introduce the Fourier representation of the Green's function components:

$$G_{\mu\nu}(\mathbf{r}, \mathbf{r}') = \frac{1}{(2\pi)^2}\int d\mathbf{k}_\parallel g_{\mu\nu}(\mathbf{k}_\parallel|zz')e^{i\mathbf{k}_\parallel \cdot (\mathbf{r}_\parallel - \mathbf{r}'_\parallel)} \tag{3.134}$$

where we have taken into account that Eq. (3.132), determining $G_{\mu\nu}$, is invariant with respect to translations along the plane $z = 0$. The zeroth-order solution $\mathbf{E}^{(0)}$ can be written in the form

$$\mathbf{E}^{(0)}(\mathbf{r}) = e^{i\mathbf{k}_{i\parallel} \cdot \mathbf{r}_\parallel}\mathbf{E}^{(0)}(\mathbf{k}_{i\parallel}|z) \tag{3.135}$$

Then the scattered field is expressed as

$$\mathbf{E}^{(1)}(\mathbf{r}) = -\frac{\omega^2}{16\pi^3 c^2}(\epsilon - 1)\int d\mathbf{k}_\parallel e^{i\mathbf{k}_\parallel \cdot \mathbf{r}_\parallel}\hat{\zeta}(\mathbf{k}_\parallel - \mathbf{k}_{i\parallel})$$

$$\times \int dz'\delta(z')\hat{\mathbf{g}}(\mathbf{k}_\parallel|zz') : \mathbf{E}^{(0)}(\mathbf{k}_{i\parallel}|z') \tag{3.136}$$

where the tensor $\hat{\mathbf{g}}$ is determined by its components $g_{\mu\nu}$ and $\hat{\zeta}(\mathbf{k}_\parallel)$ is the Fourier transform of the surface profile defined by Eq. (2.92). The integral over z' in Eq. (3.136) can be evaluated using the identity

$$\int_{-\infty}^{\infty} F(z')\delta(z')dz' = \frac{1}{2}[F(0^+) + F(0^-)] \tag{3.137}$$

and thus is expressed in terms of $g_{\mu\nu}(\mathbf{k}_\parallel|zz')$ and $E_\nu^{(0)}(\mathbf{k}_{i\parallel}|z')$ taken at $z' = 0^\pm$.

Equation (3.136) can be rewritten in the form

$$\mathbf{E}^{(1)}(\mathbf{r}) = -\frac{\omega^2}{16\pi^3 c^2}(\epsilon - 1)\int d\mathbf{G}e^{i(\mathbf{k}_{i\parallel} + \mathbf{G})\cdot\mathbf{r}_{\parallel}}\hat{\zeta}(\mathbf{G})F(\mathbf{k}_{i\parallel}, \mathbf{G}|z) \qquad (3.138)$$

with

$$F(\mathbf{k}_{i\parallel}, \mathbf{G}|z) = \int dz'\delta(z')\hat{\mathbf{g}}(\mathbf{k}_{i\parallel} + \mathbf{G}|zz') : \mathbf{E}^{(0)}(\mathbf{k}_{i\parallel}|z') \qquad (3.139)$$

which reveals that the scattered field results from the coupling of the incident wave field with the Fourier components of the surface roughness and summation over the roughness spectrum. The coupling with the Fourier-harmonic $\hat{\zeta}(\mathbf{G})$ is determined by the function $F(\mathbf{k}_{i\parallel}, \mathbf{G}|z)$ and leads to a shift of the wave vector $\mathbf{k}_{i\parallel}$ by an amount of $\mathbf{G}$.

The scattered magnetic field $\mathbf{H}^{(1)}(\mathbf{r})$ is found from the Maxwell equation $\nabla \times \mathbf{E} = -(1/c)(\partial \mathbf{H}/\partial t)$. Then one can evaluate the Poynting vector of the scattered radiation. In order to obtain an observable quantity, it should be averaged over the mean surface area A (see Section 2.1.5). This procedure involves the averaging of the product

$$\langle e^{-i(\mathbf{k}_{\parallel} - \mathbf{k}'_{\parallel})\cdot\mathbf{r}_{\parallel}}\hat{\zeta}^*(\mathbf{k}_{\parallel} - \mathbf{k}_{i\parallel})\hat{\zeta}(\mathbf{k}'_{\parallel} - \mathbf{k}_{i\parallel})\rangle$$
$$= \frac{(2\pi)^2}{A}\delta(\mathbf{k}'_{\parallel} - \mathbf{k}_{\parallel})|\hat{\zeta}(\mathbf{k}_{\parallel} - \mathbf{k}_{i\parallel})|^2 \qquad (3.140)$$
$$= (2\pi)^2\delta(\mathbf{k}'_{\parallel} - \mathbf{k}_{\parallel})P(\mathbf{k}_{\parallel} - \mathbf{k}_{i\parallel})$$

where $P(\mathbf{k}_{\parallel})$ is the power spectrum of a rough surface determined by Eq. (2.91). In the case of an isotropic surface it is reasonable to assume that the power spectrum depends on the modulus of the $\mathbf{k}_{\parallel}$-vector, but not on its direction.

The further discussion depends on which problem we are interested in. If we consider the scattering of light by a rough surface, the coordinate z in the equation for $\mathbf{E}^{(1)}$ must be taken as positive, whereas when considering the absorption of light by a surface it should be negative.

Scattering of Light
The averaged Poynting vector determines the cross-section for scattering of radiation into a unit solid angle about the direction specified by the Euler angles ϕ_s and θ_s (Fig. 3.10). The result is represented in terms of the contributions associated with the scattering of incident radiation of a given polarization into radiation with prescribed polarization. It can be written as

$$\frac{df(\mathbf{k}_{\parallel}|\alpha \rightarrow \beta)}{d\Omega_s} = \frac{\omega^4|\epsilon - 1|^2}{\pi^2 c^4}P(|\mathbf{k}_{\parallel} - \mathbf{k}_{i\parallel}|)\cos\theta_i\cos^2\theta_s|\Phi(\alpha \rightarrow \beta)|^2$$

$$\alpha, \beta = s, p \quad (3.141)$$

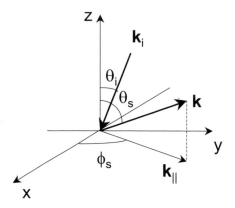

Fig. 3.10 Scattering geometry.

where

$$\Phi(s \rightarrow p) = \frac{\sin \phi_s |\epsilon - \sin^2 \theta_s|^{1/2}}{[\epsilon \cos \theta_s + (\epsilon - \sin^2 \theta_s)^{1/2}][\cos \theta_i + (\epsilon - \sin^2 \theta_i)^{1/2}]} \qquad (3.142)$$

$$\Phi(s \rightarrow s) = \frac{\cos \phi_s}{[\cos \theta_s + (\epsilon - \sin^2 \theta_s)^{1/2}][\cos \theta_i + (\epsilon - \sin^2 \theta_i)^{1/2}]} \qquad (3.143)$$

$$\Phi(p \rightarrow s) = \frac{\sin \phi_s (\epsilon - \sin^2 \theta_i)^{1/2}}{[\cos \theta_s + (\epsilon - \sin^2 \theta_s)^{1/2}][\epsilon \cos \theta_i + (\epsilon - \sin^2 \theta_i)^{1/2}]} \qquad (3.144)$$

$$\Phi(p \rightarrow p) = \frac{\cos \phi_s (\epsilon - \sin^2 \theta_s)^{1/2} (\epsilon - \sin^2 \theta_i)^{1/2} - (1/2) \sin \theta_i \sin \theta_s (\epsilon^2 + 1)}{[\epsilon \cos \theta_s + (\epsilon - \sin^2 \theta_s)^{1/2}][\epsilon \cos \theta_i + (\epsilon - \sin^2 \theta_i)^{1/2}]}$$
$$(3.145)$$

with θ_i the incidence angle.

One can see from here that the scattering cross-sections are determined by the power spectrum of the surface roughness. The nonzero $s \rightarrow p$ and $p \rightarrow s$ components indicate that the scattering at a rough surface is accompanied by *depolarization*. However, for scattering in the plane of incidence ($\phi_s = 0$) $\Phi(s \rightarrow p) = \Phi(p \rightarrow s) = 0$, i.e., within this plane the polarization of radiation is not changed. The corresponding polar diagrams for both s- and p-polarizations are shown in Fig. 3.11.

Absorption of Light.
The fraction of the energy, f_z, of an incident wave absorbed by the roughness-induced energy flow in a direction normal to the surface can be found in terms of the spatially averaged z-component of the Poynting vector $\mathbf{S}(z)$ as

$$f_z = \frac{\langle S_z(0-)\rangle}{S_{iz}} \qquad (3.146)$$

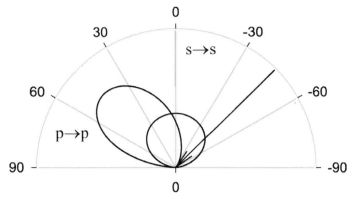

Fig. 3.11 Cross-sections of light scattering in the plane of incidence ($\phi_s = 0$) for $\epsilon = 2.25$ and $\theta_i = -45°$, and for different polarizations.

where $\mathbf{S}_i$ is the Poynting vector of the incident electromagnetic wave. In the case of a s-polarized incident wave this quantity has the form

$$f_z^s = \frac{\omega|\epsilon - 1|^2 \cos\theta_i}{\pi^2 c|(\epsilon - \sin^2\theta_i)^{1/2} - \cos^2\theta_i|^2} \int d\mathbf{k}_\parallel P(|\mathbf{k}_\parallel - \mathbf{k}_{i\parallel}|)\, \text{Re}(k_1)$$

$$\times \left(\frac{c^2}{\omega^2} \sin^2\phi_s \frac{(k_\parallel^2 + |k_1|^2)|k_z|^2}{|k_1 - \epsilon k_z|^2} + \frac{\omega^2}{c^2} \frac{\cos^2\phi_s}{|k_1 - k_z|^2} \right) \tag{3.147}$$

where $k_1 = -[\epsilon(\omega^2/c^2) - k_\parallel^2]^{1/2}$ and $k_z = [(\omega^2/c^2) - k_\parallel^2]^{1/2}$. Here, the first term in the brackets describes the $s \rightarrow p$ scattering process, while the second one arises from the scattering without change of polarization. The denominator in the first term can be rearranged to be

$$\frac{1}{k_1 - \epsilon k_z} = \frac{k_1 + \epsilon k_z}{k_1^2 - \epsilon^2 k_z^2} = \frac{k_1 + \epsilon k_z}{\epsilon^2 - 1}\left(k_\parallel^2 - \frac{\omega^2}{c^2}\frac{\epsilon}{\epsilon + 1}\right)^{-1} \tag{3.148}$$

Therefore, if $\text{Im}(\epsilon)$ is small [21] and $\text{Re}(\epsilon) < 0$, the integrand in Eq. (3.147) has a sharp maximum at $k_\parallel = q$ with

$$q = \frac{\omega}{c}\sqrt{\frac{\epsilon}{\epsilon + 1}} \tag{3.149}$$

the wave vector of a surface polariton (see Section 3.3.1). Introducing the new integration variable, $\mathbf{G} = \mathbf{k}_\parallel - \mathbf{k}_{i\parallel}$, one can rewrite the integral corresponding to this term in the form

$$\int d\mathbf{G} \frac{P(G)}{|\mathbf{k}_{i\parallel} + \mathbf{G}|^2 - q^2} H(\mathbf{G}) \tag{3.150}$$

where $H(\mathbf{G})$ is a slowly varying function of $\mathbf{G}$. The resonance in the integrand at

$$|\mathbf{k}_{i\parallel} + \mathbf{G}| = q \tag{3.151}$$

21) Note that the left-hand side of Eq. (3.148) can be equal to zero only when $\text{Re}(\epsilon) < 0$.

corresponds to the excitation of surface polariton through the scattering of the incident light at the Gth Fourier-harmonic of the surface roughness.[22] The strength of this coupling, and hence the absorbed energy, is proportional to the Gth component of its power spectrum, $P(G)$.

The integral over $k_\parallel$ in Eq. (3.147) can be approximately evaluated by taking all slowly varying functions at $k_\parallel = q$. The remaining integration is then reduced to the integral over ϕ_s and the result for $\text{Im}(\epsilon) \to 0$ has the form

$$f_z^s = \frac{\omega^4}{c^4} \frac{\epsilon^2 \cos \theta_i}{(\epsilon^2 - 1)^{5/2}} \int_0^{2\pi} \frac{d\phi_s}{\pi} \sin^2 \phi_s P(|k_{sp}\hat{\mathbf{n}} - \mathbf{k}_{i\parallel}|) \tag{3.152}$$

where $\hat{\mathbf{n}}$ is a unit vector along the vector $\mathbf{k}_\parallel$.

Under the same assumptions the fraction of the energy, $f_\parallel^s$, of the incident wave absorbed by the energy flow parallel to the surface is given by

$$|f_\parallel^s| = \frac{\omega^4}{c^4} \frac{\epsilon^2 \cos \theta_i}{(\epsilon - 1)^{5/2}} \frac{L}{L_\parallel} \int_0^{2\pi} \frac{d\phi_s}{\pi} \cos \phi_s \sin^2 \phi_s P(|k_{sp}\hat{\mathbf{n}} - \mathbf{k}_{i\parallel}|) \tag{3.153}$$

where L is the propagation length of the surface polariton and $L_\parallel$ is the linear dimension of the surface area illuminated by the incident beam along its wave vector component $\mathbf{k}_{i\parallel}$.

Further Reading

V.M. Agranovich, D.L. Mills (Eds.), *Surface Polaritons*, North-Holland Publishing Company, Amsterdam, 1982.

M. Born and E. Wolf, *Principles of Optics*, Pergamon Press, Oxford, 1975.

L.D. Landau and E.M. Lifshitz, *Electrodynamics of Continuous Media*, Pergamon Press, Oxford, 1963.

J.A. Ogilvy, *Theory of Wave Scattering from Random Rough Surfaces*, Adam Hilger, Bristol, 1991.

H. Raether, *Surface Plasmons on Smooth and Rough Surfaces and on Gratings*, Springer-Verlag, Berlin, 1988.

22) Compare with Eq. (3.109).

Problems

Problem 3.1. Find the reflectivity of p-polarized light incident from a vacuum at a weakly absorbing medium (the dielectric function $\epsilon_2 = \epsilon_2' + i\epsilon_2''$ with $|\epsilon_2''| \ll \epsilon_2'$) under the Brewster angle $\theta_B = \arctan\sqrt{\epsilon_2'}$.

Problem 3.2. The surface of a transparent crystal (index of refraction n) is covered with a weakly absorbing film (dielectric function $\epsilon_2 = \epsilon_2' + i\epsilon_2''$ with $|\epsilon_2''| \ll \epsilon_2'$) of thickness d. Light of wavelength $\lambda \gg d$ falls from the crystal side at an angle θ_i (with respect to the surface normal) which is greater than the critical angle at the crystal–vacuum interface. Find the reflectivity of light for s- and p-polarizations.

Problem 3.3. The surface of a transparent crystal (index of refraction n) is covered with a film (dielectric function ϵ_2, the thickness $d \ll \lambda$). Find the reflection coefficient of p-polarized light incident at the film from the vacuum side under the Brewster angle.

Problem 3.4. Prove that Eqs (3.57) and (3.58) take the form of Eqs (3.44) and (3.45) if the transition layer is macroscopic and its optical properties can be characterized by a dielectric function.

Problem 3.5. Prove that, if an interface between two media can be described by a transient layer with effective isotropic dielectric function and effective thickness, Eqs (3.75) and (3.76) derived for nonlocal optical response are reduced to Eqs (3.44) and (3.45), respectively.

Problem 3.6. Prove that surface polaritons cannot have a TE polarization.

Problem 3.7. Neglecting relaxation, find the penetration depths of surface plasmon polaritons into both metal and vacuum for wavelengths λ much longer than the plasma wavelength, λ_p. Calculate them for a surface plasmon polariton at an Al surface ($\lambda_p \approx 800$ Å) with $\lambda = 5890$ Å.

Problem 3.8. Neglecting relaxation, obtain the dispersion relation of surface plasmon polariton at an interface between two metals having the plasma frequencies ω_{p1} and ω_{p2}. In what frequency range can a surface plasmon polariton exist? Find its frequency in the nonretarded limit ($c \to \infty$).

Problem 3.9. Determine the frequency range in which surface polaritons at a surface of an n-type semiconductor can be excited. Take the following values of the parameters: $\omega_{TO} = 180$ cm^{-1}, $\epsilon_0 = 9.3$, $\epsilon_\infty = 1.9$, $m^* = 0.02\, m_e$ and $n = 2 \times 10^{17}$ cm^{-3}.

Note: Use a graphical solution method.

Problem 3.10. A light beam of wavelength $\lambda = 590$ nm falls from a vacuum onto a metal surface with a sinusoidal grating of period $d = 1.77$ µm. The dielectric function of the metal is $\epsilon(\lambda) = -9$. Find all possible angles of incidence at which light will excite surface plasmon polaritons, provided that the plane of incidence is perpendicular to the grating ruling.

Problem 3.11. Calculate the propagation length of a surface polariton excited at $\lambda = 6000$ Å if the ATR minimum in a Kretschmann configuration at the angle of incidence $\theta_m = 45°$ equals zero and its full width at half minimum is $\Delta\theta = 0.5°$. The index of refraction of the prism is $n = 1.5$.

4
Infrared Spectroscopy at Surfaces and Interfaces

The frequencies of vibrational motion in crystals and molecules fall into the infrared (IR) spectral range. IR radiation illuminating the sample surface can therefore excite either phonons in crystals or vibrations of adsorbed molecules. Their excitation is most efficient when the frequency of the IR radiation is close to the internal vibrational frequencies of the sample. As a result, the optical response is a maximum at resonance and decreases when the detuning from resonance increases. This feature allows one to determine the vibrational frequencies or to identify molecules present at the surface if their frequencies are known.

IR radiation occupies the spectral range between the red edge of visible light ($\lambda \sim 0.76$ μm) and the short radio waves ($\lambda \approx 1$–2 mm). This dictates the use of specific IR sources and IR detectors. Traditionally, a rod heated up to a high temperature has been used to generate the incident IR radiation.[1] Recently, however, new bright sources of IR light such as IR lasers, free electron lasers and synchrotron radiation from electron storage rings have become available. Many of them can provide both polarized and frequency-tunable light. In view of spectroscopic applications this is an essential advantage. IR detectors can be broadly classified as thermal and photoelectrical. The former ones, bolometers, operate on the principle that IR absorption in a sensitive element leads to heat generation and an increase of the internal resistance. In the latter ones, the absorption of IR photons induces an electric current or a voltage in the detector.

In this chapter we shall consider various experimental techniques based on application of IR radiation which are commonly used for surface and interface analysis.

---

1) Such a homogeneous "blackbody" radiator at constant temperature over the whole emitting area is called a Globar.

Optics and Spectroscopy at Surfaces and Interfaces. Vladimir G. Bordo and Horst-Günter Rubahn
Copyright © 2005 WILEY-VCH Verlag GmbH & Co. KGaA, Weinheim
ISBN: 3-527-40560-7

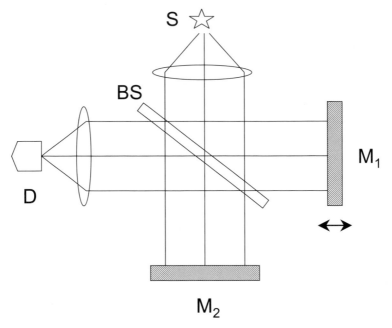

Fig. 4.1 Schematic drawing of a Michelson interferometer. S, source; BS, beam splitter; M_1, moving mirror; M_2, fixed mirror; D, detector.

4.1
Infrared Spectroscopic Ellipsometry (IRSE)

As we have seen in Section 3.1.1, the ratio of the reflection coefficients in p- and s-polarizations, ρ, can be represented in terms of the ellipsometric angles Ψ and Δ (Eq. (3.33)). These quantities are determined by the sample properties and being measured as a function of the incident radiation frequency provide spectroscopic information about the sample. This is the basic idea of spectroscopic ellipsometry or, in the IR region, *infrared spectroscopic ellipsometry* (IRSE).

For spectral measurements with a high signal-to-noise ratio one usually uses a Fourier transform infrared (FTIR) spectrometer. It is based on a Michelson interferometer where one of the mirrors is movable (Fig. 4.1). The path difference, l, is changed continuously by moving the mirror M_1. As a result, the interferogram is a function of l:

$$I(l, \bar{v}) = I_0 \cos^2(\pi \bar{v} l) = \frac{I_0}{2} + \frac{I_0}{2} \cos(2\pi \bar{v} l) \qquad (4.1)$$

where $\bar{v} = 1/\lambda$ is the wavenumber of the incident radiation. For a continuous spectrum confined between $\bar{v}_{min}$ and $\bar{v}_{max}$ and described by the function $I(\bar{v})$

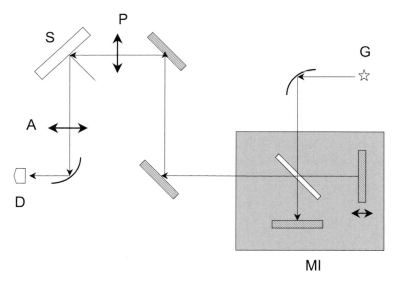

Fig. 4.2 An FTIR ellipsometer. MI, Michelson interferometer; G, globar source; S, sample; P, polarizer; A, analyzer; D, detector.

the varying part of the interferogram is determined by the integral

$$I(l) = \frac{1}{2} \int_{\bar{\nu}_{\min}}^{\bar{\nu}_{\max}} I(\bar{\nu}) \cos(2\pi\bar{\nu}l) d\bar{\nu} \qquad (4.2)$$

Then the frequency spectrum, $I(\bar{\nu})$, is obtained by a Fourier transform of Eq. (4.2):

$$I(\bar{\nu}) = 2 \int_{0}^{l_{\max}} I(l) \cos(2\pi\bar{\nu}l) dl \qquad (4.3)$$

where $l_{\max}$ is the maximum path difference.

A FTIR spectrometer is the main part of a FTIR ellipsometer which is schematically shown in Fig. 4.2. This device is commonly used for spectral investigations of thin solid films and surface overlayers in the IR region. Whereas ellipsometry with visible and ultraviolet light (Section 5.1) is mainly constrained to study the surface electronic properties, IRSE also provides information about surface structures. The vibrational bands, which are probed by IRSE, represent fingerprints of the compounds under consideration and depend on the geometrical arrangement of the molecules in the film. This allows one to deduce the chemical composition of the sample as well as its crystal structure. In addition, a large range of film thicknesses from a few nanometers to several micrometers can be examined.

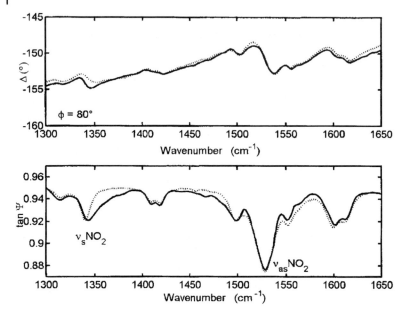

Fig. 4.3 Infrared ellipsometric spectra of a Langmuir–Blodgett film of 2,5-diphenyl-1,3,4-oxadiazole deposited on a gold surface. The incidence angle of the light is 80°. The dotted curves are the calculated spectra. Reprinted with permission from (Tsankov et al. 2001). Copyright 2001, Wiley InterScience.

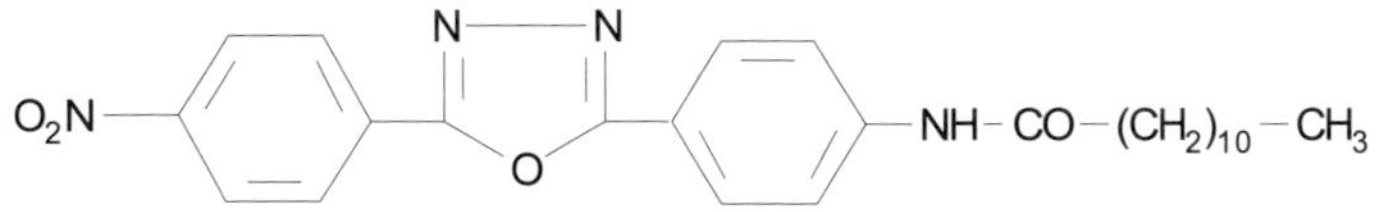

Fig. 4.4 Chemical structure of the 2,5-diphenyl-1,3,4-oxadiazole.

Figure 4.3 illustrates the application of the IRSE technique for the study of a Langmuir–Blodgett (LB) film[2] consisting of nine double layers of 2,5-diphenyl-1,3,4-oxadiazole (Fig. 4.4) deposited on a gold surface (Tsankov et al. 2001). Since the head NO_2 groups in the LB film are oriented towards the surface, the dipole moment of the symmetric stretch vibration ν_s (NO_2) at 1342 cm^{-1} is perpendicular to the surface, while that of the asymmetric vibration ν_{as} (NO_2) at 1526 cm^{-1} is almost parallel to it. Therefore, taking into account the incidence angle of 80°, one would expect a notable polarization effect provided that the molecules were standing precisely upright. However, both vibrational bands are prominent in the spectrum. This finding implies that the molecules in every individual layer are tilted.

2) Langmuir–Blodgett films are constituted by individual monolayers of amphiphilic molecules consecutively deposited from a water surface onto a solid substrate.

4.2

Infrared Reflection–Absorption Spectroscopy (IRAS) of Adsorbed Molecules

IR radiation striking a surface can be absorbed if its frequency is close to the frequency ω_0 of a vibrational transition in the adsorbed molecules. As a result, the spectrum of reflected light will display a dip at the vibrational transition frequency whose depth is proportional to the surface density of adsorbed species having vibrational modes of frequency ω_0. This principle is used as the basis of *infrared reflection–absorption spectroscopy* (IRAS). The vibrational spectrum of a molecule presents its "fingerprint" and therefore the presence or absence of particular vibrational bands in an IRAS spectrum can be used for characterization of an adsorbed overlayer. The sensitivity of such measurements can be as good as 0.5% of a monolayer. Another application of IRAS is based on its high spectral resolution, typically about 0.05 meV. It allows precise measurement of frequency, width and coverage-dependent shift of vibrational bands. This data in turn gives information on adsorption sites, lateral interactions, adsorbate growth as well as on vibrational lifetimes and dephasing effects.

For electric dipole transition from an initial state $|i\rangle$ to a final state $|f\rangle$ the intensity of absorption of radiation at the frequency ω is given by

$$I(\omega) \sim |\vec{\mu}_{fi} \cdot \vec{E}|^2 g(\omega) \tag{4.4}$$

where $\vec{\mu}_{fi} \equiv \langle f|\vec{\mu}|i\rangle$ is the transition dipole moment and $\vec{E}$ is the electric field vector at the surface. The function $g(\omega)$ describes the absorption lineshape. It is centered at the transition frequency ω_0 and its width is determined to first order by the quantity $\gamma_{\perp}^s$ (see Section 2.2.4). A spread in transition frequencies of molecules because of different local environments leads to additional *inhomogeneous broadening* of the absorption/reflection spectra.

As follows from Eq. (4.4), absorption occurs if the scalar product $\vec{\mu}_{fi} \cdot \vec{E}$ is nonzero. Using an adiabatic approximation and assuming small normal vibrational amplitudes Q_j one can write the transition dipole moment in the form

$$\vec{\mu}_{fi} \approx \sum_j \left(\frac{\partial \langle \vec{\mu} \rangle}{\partial Q_j} \right)_0 \langle \phi_f | Q_j | \phi_i \rangle \tag{4.5}$$

where $\langle \vec{\mu} \rangle$ is the molecular dipole moment averaged over the ground electronic state, ϕ_i and ϕ_f are the nuclear wave functions in the initial and final states, respectively, and the partial derivative is taken at the equilibrium nuclear configuration. The condition $\vec{\mu}_{fi} = 0$ determines *selection rules* for the transition $|i\rangle \rightarrow |f\rangle$. For room temperature and below, and for $\omega_0 \geq 500\,\mathrm{cm}^{-1}$, the populations of the excited vibrational states can be neglected. Then one can substitute $|i\rangle$ with the ground vibrational state and $|f\rangle$ with the first excited vibrational state. In this case, according to group theory, a vibrational

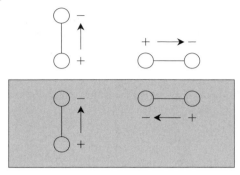

Fig. 4.5 Dynamic molecular and image dipoles oriented parallel and perpendicular to a metal surface (indicated by a shaded block).

transition can occur, i.e., it is *allowed* or *dipole-active* only if the given vibrational mode Q_j belongs to the same irreducible representation of the point symmetry group as at least one of the coordinates x', y' or z' in the molecular frame of reference.

Absorption of light by molecules adsorbed on metal surfaces requires special discussion. The incident wave induces a dynamic dipole moment $\vec{\mu}(\omega)$ in a molecule whose imaginary part determines the absorbed energy. Its interaction with the metal substrate can be described in terms of an image dipole located at the position of a mirror image of the molecular dipole (see Section 2.2.1). For a perfect conductor ($\epsilon = \infty$), the component of the molecular dipole parallel to the surface is compensated exactly by its image, whereas the component perpendicular to the surface is enhanced by a factor of two as compared to its value far from the surface (Fig. 4.5). In addition, the component of the electric field parallel to the surface is also screened by a metal substrate. The combination of the two screening effects gives the so-called *pseudo-selection rules* at a metal surface: spectral lines corresponding to vibrational modes accompanied by atomic displacements parallel to the surface are suppressed strongly in intensity (cf., Fig. 2.19). Only those originating from vibrations perpendicular to the surface are observed in IRAS. In this case, instead of Eq. (4.4) one has

$$I(\omega) \sim |\,(\mu_z)_{fi}\, E_z|^2 g(\omega) \qquad (4.6)$$

Comparison of an observed frequency shift of a molecule on adsorption with predicted values (Section 2.2.3) sometimes can help to identify the mechanism responsible for this effect. For example, for the C–O stretching vibration in the CO/Pt system, the mechanical renormalization gives a blue shift of about 50 cm^{-1} from the gas-phase value of 2143 cm^{-1}. On the other hand, the interaction of the CO molecule with its own image leads to a red shift of approximately the same amount which cancels the shift due to renormalization. In order to explain the observed decrease in the vibrational frequency of CO molecules adsorbed on metal surfaces (which reaches about 350 cm^{-1} on a Pd(111) surface) one has to invoke the effect of the chemical bonding between CO and the specific metal.

The width of a vibrational band provides additional information on adsorption. Typical bandwidths for adsorbed species on metal surfaces vary from a few cm^{-1} to several hundred cm^{-1}. Various mechanisms can contribute to them (Section 2.2.4) and their separation is sometimes not possible. The mechanisms responsible for the energy decay are not only interesting from the viewpoint of relaxation of excited states at the surface. The same decay channels determine the transfer of kinetic energy from the incoming particle into elementary excitations of the substrate and hence govern surface dynamics.

The main experimental problem in IRAS is the small number of particles under investigation, which means that high sensitivity is required. As follows from Eq. (4.6), the signal in IRAS is proportional to $|E_z|^2$. This quantity is nonzero when the incident wave is p-polarized and it increases with the angle of incidence, θ_i. Taking into account that the surface area illuminated by the incident beam varies as $1/\cos\theta_i$, one usually plots $|E_z|^2/\cos\theta_i$ as a function of θ_i. This dependence has a maximum between 80 and 90°, thus determining the optimum measurement geometry. In contrast to metal surfaces, the maximum reflectivity change in IRAS at nonmetal surfaces takes place at the Brewster angle. Here the reflectivity of the substrate is lowest, or even zero (for p-polarization). Thus, even very small perturbations due to deposition of an adsorbate can cause an essential change in reflectivity. Additional factors, among which are commensurability between the beam diameter and the sample dimensions, the use of focusing optics and the signal-to-noise ratio, can influence the optimum experimental conditions for IRAS measurements. Further gain in sensitivity can be obtained with FTIR spectrometers.

Figure 4.6 shows the IRAS spectra of the CO/Pt system measured with a spectral resolution of 2 cm^{-1} using a FTIR spectrometer. These spectra allow one to follow the adsorption of CO molecules with increasing surface coverage, θ. Simultaneously performed low-energy electron diffraction measurements which essentially provide structural information, indicate that a single band at ca. 2100 cm^{-1} is due to the C–O stretch vibrations at on-top sites, whereas that at 1850 cm^{-1} originates from bridge sites. The spectra thus reveal that the occupation of the bridge sites begins at $\theta = 0.32$. The on-top vibrational band shifts from 2089 to 2103 cm^{-1} when the coverage increases up to $\theta = 0.5$. This shift can be attributed to lateral interactions between adsorbed CO molecules.

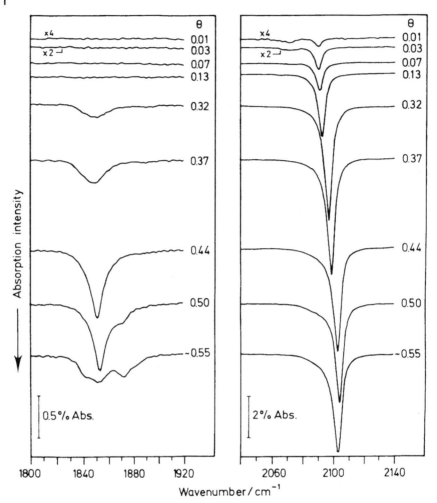

Fig. 4.6 IRAS spectra of the C–O stretching vibration in the CO/Pt(111) system as a function of surface coverage at 125 K. Reprinted from (Tüshaus et al. 1987), Copyright 1987, with permission from Elsevier.

4.3
Infrared Surface Polariton Spectroscopy

In Section 3.3 we have considered surface polaritons which are surface electromagnetic excitations propagating along a surface or an interface. Their amplitude has a maximum at the boundary between two contacting media. This feature leads to a strong influence of the surface/interface properties on the SP characteristics. SPs excited in the IR spectral region are, therefore, sensitive optical probes of vibrations in both the substrate crystal and the overlayer. On

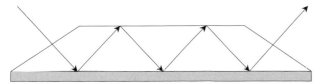

Fig. 4.7 Schematic diagram of the multiple ATR method, showing an absorbing layer and a Dove prism on top of it.

metal surfaces, infrared SPs can propagate over macroscopic distances of the order of a few centimeters. This provides additional gain in sensitivity.

To investigate the SP dispersion law in the IR region one usually exploits the attenuated total reflection (ATR) method. In practice, a more convenient way of measurements is scanning over the incident beam frequency while keeping the angle of incidence fixed. Exact determination of the incidence angle requires a small angular divergence of the light beam. For these purposes, tunable lasers are ideal sources of IR radiation. As the prism material various crystals and glasses are used which are transparent in the IR spectral region (Si, CaF_2, KRS-5, etc.). To further increase the sensitivity of the method, one can apply multiple total internal reflections (Fig. 4.7).

The dispersion law which dictates the conditions of SP excitation is determined by the optical constants of the bordering media. On the other hand, the presence of a film or an adsorbed layer at the interface modifies the dispersion relation and thus allows one to extract information on optical properties of an overlayer. These two approaches are used as the basis of various experimental techniques.

4.3.1
Determination of Optical Constants of Crystals

In Section 3.3.3 we have considered surface phonon polaritons in dielectrics and semiconductors. To derive their dispersion relation we have used the dielectric function of a crystal in the form (3.96). However, for comparison with experimental data it is necessary to take into account the decay rate of phonons, $\Gamma(\omega)$, as well as their anharmonicity characterized by the frequency shift $\Delta(\omega)$ (Mirlin 1982). The corresponding dielectric function is given by

$$\epsilon(\omega) = \epsilon_\infty + (\epsilon_0 - \epsilon_\infty) \frac{\omega_0^2 + 2\Delta(\omega)\omega_0}{\omega_0^2 - \omega^2 + 2[\Delta(\omega) + i\Gamma(\omega)]\omega_0} \tag{4.7}$$

with ω_0 the frequency of TO-phonons in the harmonic approximation.

Although there are many methods available to determine $\epsilon(\omega)$ for a crystal from spectra of transmissivity, external reflection, ellipsometry, etc., they are in general nonversatile and useful only for a limited range of optical constants. The advantage of surface polariton spectroscopy is that it provides a direct

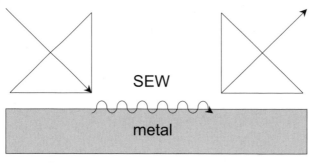

Fig. 4.8 Schematic of the two-prism method for the study of surface electromagnetic wave propagation.

measurement of the surface phonon polariton frequency from the position of the ATR minimum as well as its decay at this frequency (from the width of the minimum). Comparison of experimental spectra with those calculated using Eq. (4.7) allows one to determine the frequency dependencies of both $\Gamma(\omega)$ and $\Delta(\omega)$. Then, from the measured dielectric function, one can obtain the index of refraction, $n(\omega)$, and the absorption index, $\kappa(\omega)$, using the equation

$$\epsilon(\omega) \equiv \epsilon'(\omega) + i\epsilon''(\omega) \equiv [n(\omega) + i\kappa(\omega)]^2 \qquad (4.8)$$

On metal surfaces, surface plasmon polaritons in the IR range run to macroscopic distances and they are essentially *surface electromagnetic waves* (SEWs) (Zhizhin et al. 1982). To study their propagation one uses a two-prism method (Fig. 4.8). The first prism transforms the incident light into a SEW, whereas the second one performs the reverse transformation. The dielectric function of a metal, considering scattering of conduction electrons, has the form

$$\epsilon(\omega) = 1 - \frac{\omega_p^2}{\omega(\omega + i\Gamma)} \qquad (4.9)$$

with ω_p the plasma frequency of the metal, Eq. (2.65), and Γ the rate of electron collisions with phonons, electrons and impurities. Then, with $\omega \gg \Gamma$, the propagation length of a SEW on a metal surface is given by[3]

$$L \approx \frac{\omega_p^2 c}{\omega^2 \Gamma} \qquad (4.10)$$

Equation (4.10) allows one to determine Γ from measurements of the propagation length if ω_p is known.

---

3) We have taken into account here that in the IR range $|\operatorname{Re}(\epsilon)| \gg 1$.

4.3.2
Surface Electromagnetic Wave Spectroscopy of Overlayers

If a crystal surface is covered with a film of thickness d much smaller than the wavelength of the incident radiation, the absorption coefficient, $\alpha(\omega)$,[4] of the surface polariton can be written as (Zhizhin et al. 1982)

$$\alpha(\omega) \approx \alpha_0(\omega) + \Delta\alpha(\omega) \qquad (4.11)$$

where α_0 is the absorption coefficient at the interface without the film and $\Delta\alpha$ is the contribution of the film in absorption. If the film absorbs selectively, $\Delta\alpha(\omega)$ has resonances at its vibrational frequencies. For a metal substrate the absorption maximum at the frequency of longitudinal vibrations of the film is much greater than that at the frequency of transverse vibrations. In this case

$$\Delta\alpha(\omega) \approx 2d \, \frac{\omega^2}{c^2} \frac{\Gamma}{\omega_p} \, \mathrm{Im} \left[-\frac{1}{\epsilon_f(\omega)} \right] \qquad (4.12)$$

where ϵ_f is the dielectric function of the film.

The absorption spectrum of thin films can most conveniently be obtained by a differential method. Here, one measures the ratio of the SEW intensity for a clean surface, I_0, to that for a surface covered with a film, I, at different distances between the prisms, r. Then the SEW absorption is determined as

$$\Delta\alpha = \frac{1}{L} - \frac{1}{L_0} = \frac{1}{r} \ln \frac{I_0}{I} \qquad (4.13)$$

where L_0 and L are the SEW propagation lengths on a clean surface and on a surface with a film, respectively.

Figure 4.9 shows the SEW absorption spectrum of the Langmuir–Blodgett films of sim-di(γ-carboxydecyl)tetramethylsiloxane on a Cu surface measured by this method. Both spectra, obtained for one monolayer and for three monolayers deposited on the surface, have an absorption band with a maximum at $1075\,\mathrm{cm}^{-1}$ corresponding to a valent vibration of the group Si-O-Si. This maximum neither shifts nor broadens with transition from one to eleven monolayers, suggesting that the vibration does not interact strongly with the substrate or with the vibrations in the other monolayers. The absolute error of differential absorption in these measurements was $\pm 0.005\,\mathrm{cm}^{-1}$.

4.3.3
Correlation between Propagation Length and Surface Roughness

On an ideally flat metal surface the SEW absorption is related to the electron collisions with phonons, impurities and other electrons. Surface roughness introduces an additional decay path for the SEWs. The reason for this is twofold.

---

4) The absorption coefficient is defined as $\alpha = 2\,\mathrm{Im}\,q$ with q the wave vector of surface polariton.

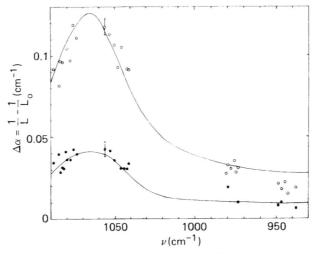

Fig. 4.9 SEW absorption spectrum of one (filled circles) and three (open circles) monolayers of sim-di(γ-carboxydecyl)tetramethylsiloxane on a Cu surface. The calculated data are shown by solid lines. Reprinted from (Zhizhin et al. 1980), Copyright 1980, with permission from Elsevier.

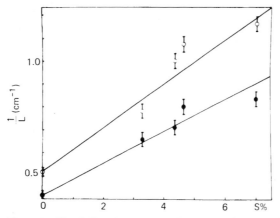

Fig. 4.10 The SEW absorption coefficient as a function of the total area of defects on Cu surfaces at 941 cm^{-1} (filled circles) and at 1041 cm^{-1} (open circles). Reprinted with permission from (Zhizhin et al. 1982).

First, the mean free path of conduction electrons is reduced due to scattering on surface defects. Secondly, the SEW is scattered on surface roughness, resulting in emission of light (Section 3.3.4). This correlation can be used, for example, to control the quality of metal mirrors (Zhizhin et al. 1982).

Figure 4.10 shows the results of measurements of the SEW propagation length on copper mirrors with different total area of surface defects. The mean height of the surface roughness of the samples was between 370 and 580 Å.

The SEW absorption increased approximately linearly with increasing area of defects. Numerical estimates indicated that the dominant contribution to the decay of the SEW intensity originates from the electron scattering on these surface defects.

4.4
Time-resolved Infrared Spectroscopy

Vibrational spectroscopy of adsorbates enables one to identify in a static manner the molecular composition and adsorbate interactions in adlayers. The development of ultrafast laser techniques opens up the possibility carrying out time-resolved measurements on picosecond and femtosecond timescales and thus to investigate dynamical aspects of adsorption. In particular, time-domain measurements of excited-state relaxation phenomena allow one to study vibrational energy transfer at surfaces and interfaces (Cavanagh et al. 1994). Such measurements provide specific information which cannot be obtained from a vibrational lineshape analysis. The vibrational linewidth is determined by both homogeneous and inhomogeneous broadenings. The vibrational-state lifetime contributes only to the former mechanism. Even for a homogeneous sample it is difficult to distinguish between the contributions from dephasing and vibrational energy decay processes. In addition, determination of energy decay pathways from linewidths is a challenging task. These problems vanish in time-resolved experiments which identify particular energy-transfer channels and measure their rates.

Such experiments are usually based on pump-probe schemes where an intense picosecond IR pulse, resonant to a vibrational transition in an adsorbate, creates a nonequilibrium population of an excited state. The subsequent evolution of this population is probed with a second, time-delayed IR pulse. The first time-resolved measurements of adsorbate vibrational relaxation were carried out for hydroxyl groups bound to colloidal SiO_2 (Heilweil et al. 1984). The use of colloidal particles of about 10 nm in diameter leads to an increase in the effective number of adsorbate monolayers up to 10^5. The frequency of the vibrational transition $v = 0 \rightarrow v = 1$ of surface hydroxyl groups is much higher than the frequencies of the substrate phonon modes. This allows one to monitor the evolution of the excited vibrational state population in the transmission of IR radiation. Due to the anharmonicity of the vibrations, absorption of the pump IR light does not lead to transitions to higher vibrational levels with $v > 1$ (see Fig. 4.11). The pumping of the $v = 0 \rightarrow v = 1$ transition, therefore, causes a transient decrease in the population difference between the levels $v = 0$ and $v = 1$ and, accordingly, a transient increase in sample transmission, T, at the transition frequency. As the population relaxes, the sample transmission returns to its equilibrium value, T_0, which is monitored by a weak,

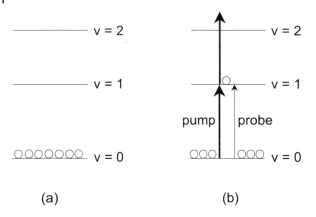

(a)   (b)

Fig. 4.11 Scheme of vibrational levels of an adsorbed OH molecule. (a) Equilibrium populations of vibrational levels. Due to the large vibrational quantum, only the ground level is significantly populated. (b) The pump pulse populates the excited level $v = 1$ and does not populate the higher levels because of anharmonicity. The probe pulse measures transient changes in sample transmission.

time t_D delayed probe pulse at the same frequency. The data are represented as a plot of $\ln(T/T_0)$ versus t_D and a linear fit provides the vibrational relaxation rate. It is worthwhile to note that this technique allows one to measure the vibrational energy relaxation time of an adsorbate surrounded by different solvents (Heilweil et al. 1985).

A slightly modified method can be applied if the investigated molecules are adsorbed on a nontransparent substrate. In this case the vibrational relaxation of molecules excited by an IR pulse can be monitored by the recovery of the surface absorption to its equilibrium value. The transient change in absorption is determined from the relative intensities of the reflected p and s components of the probe beam, I_p/I_s. This technique was used, for example, to measure the transient response of vibrationally excited CO molecules adsorbed on a Pt(111) surface (Beckerle et al. 1990). Alternatively, one can probe the vibrational relaxation of excited admolecules by the use of nonlinear optical techniques which will be considered in Chapter 6.

More substantial information on the vibrational energy transfer in an adsorbate can be obtained from transient IR *spectra* associated with a vibrationally excited adsorbate. The spectral resolution (< 1 cm^{-1}) is achieved by introduction of a grating infrared spectrometer in the detection optics following the pump-probe interaction at the sample (Fig. 4.12). Transient absorption is observed by measuring the relative intensity of the probe beam reflected from the sample, I_p/I_s, and normalizing it to the same intensity ratio in the absence of the pump, $(I_p/I_s)_0$. The signal is then defined as

$$-\Delta\alpha = \ln(I_p/I_s) - \ln(I_p/I_s)_0 \qquad (4.14)$$

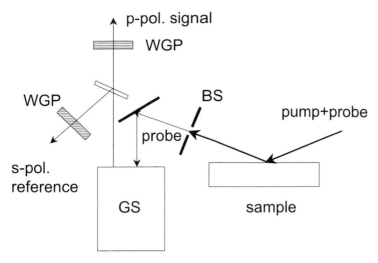

Fig. 4.12 Schematic of the experimental setup for measurements of spectrally re-solved vibrational spectra: BS, beam stop; GS, IR grating spectrometer; WGP, wire grid polarizer. The pump beam is p polarized; the polarization of the probe beam is rotated approximately 45° out of the plane of incidence.

This setup allows one to obtain two types of transient data: (i) at fixed t_D, the monochromator grating is scanned; or (ii) at fixed monochromator frequency, the time delay is changed. Figure 4.13 shows the transient absorption spectra of CO molecules adsorbed on a Pt(111) surface and obtained for different time delays. A distinct shift to lower frequencies and a spectral broadening are observed at small t_D as well as the rapid return of the spectrum to the equilibrium lineshape. Such temporal behavior can be described as originating from a one-phonon band of the adlayer which shifts with the increasing excited-state population.

Besides the decay of vibrational excitation of an adsorbate, time-resolved measurements permit one to study the reverse process, namely energy transfer from the excited carriers and lattice vibrations in the substrate to the molecular adlayer. In these experiments the picosecond pump pulse, not resonant with any molecular vibrational transition, creates a transient increase in the near-surface electron temperature. The nonequilibrium electrons can then equilibrate either with bulk phonon modes leading to surface heating or directly with the adsorbate vibrational modes. Then the transient population of the adsorbate excited state can be monitored by picosecond IRAS. For example, in the experiments with the CO/Pt(111) system, excitation of the substrate with visible laser pulses is accompanied by an increase in the electron temperature of $\Delta T \approx 80$ K. This leads to a significant excitation of solely the low-frequency frustrated translational mode ($\nu_4 = 60$ cm^{-1}). Although the signal-to-noise ratio and the strong electron–phonon coupling did not allow

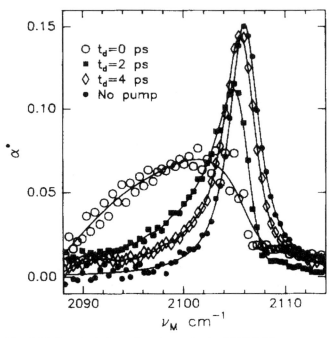

Fig. 4.13 Transient absorption spectra for the CO/Pt(111) system as a function of the time delay between the probe and pump pulses. Reprinted from (Cavanagh et al. 1994), Copyright 1994, with permission from Elsevier.

one to distinguish between the two possible energy transfer channels (viz., bulk phonon vs. adsorbate modes), the measurements established the limits for the coupling through both mechanisms (Germer et al. 1993).

Further Reading

V.M. Agranovich, D.L. Mills (Eds.), *Surface Polaritons*, North-Holland Publishing Company, Amsterdam, 1982.

R.M. Assam, N.M. Bashara, *Ellipsometry and Polarised Light*, North-Holland Publishing Company, Amsterdam, 1997.

A.M. Bradshaw, E. Schweizer, Infrared Reflection–Absorption Spectroscopy of Adsorbed Molecules in *Advances in Spectroscopy: Spectroscopy of Surfaces*, Vol. 16, R.J.H. Clark, R.E. Hester (Eds.), John Wiley & Sons, Chichester, 1988.

R.R. Cavanagh, E.J. Heilweil, J.C. Stephenson, Time-resolved measurements of energy transfer at surfaces, *Surf. Sci.* **1994**, *299/300*, 643.

Y.J. Chabal, Surface infrared spectroscopy, *Surf. Sci. Rep.* **1988**, *8*, 211.

A. Röseler, *Infrared Spectroscopic Ellipsometry*, Akademie Verlag, Berlin, 1990.

Problems

Problem 4.1. A metal film of thickness d (dielectric function $\epsilon_2 = \epsilon'_2 + i\epsilon''_2$) is deposited onto the surface of a transparent crystal (index of refraction $n \ll \sqrt{|\epsilon'_2|}$). Determine the ellipsometric angles for light of wavelength $\lambda \gg d$ incident from a vacuum under the Brewster angle.

Problem 4.2. In the IRAS spectrum from the W(100)−H system one observes a single line corresponding to the normal vibrational mode of hydrogen. From which mode does it originate? Why are the other modes not displayed in the spectrum?

Problem 4.3. One observes two lines in the IRAS spectrum obtained with p-polarization from the Si(100)−H system. To which normal modes do they correspond? Which line will disappear after excitation with s-polarized light? Can these lines originate from atoms adsorbed at: (a) a hollow site? (b) an on-top site? (c) a bridge site?

Problem 4.4. Obtain the condition under which a surface polariton at a crystal–vacuum interface propagates over macroscopic distances ($L \gg \lambda$). Express the dispersion relation of the SP and its propagation length in terms of the dielectric function of a crystal in this case.

Problem 4.5. Neglecting anharmonicity of phonons, obtain an expression for the propagation length of surface phonon polariton at a crystal–vacuum interface at frequencies close to ω_{TO} in terms of the decay rate of phonons, $\Gamma \ll \omega_{TO}$.

5

Linear Optical Techniques at Surfaces and Interfaces

The frequencies of electronic transitions, both in crystals and in adsorbed atoms or molecules, lie in or around the visible region of the spectrum. Therefore they can be excited with various sources of visible or ultraviolet light. A lamp combined with a monochromator is the simplest device which allows one to measure an optical spectrum. However, nowadays bright light sources such as tunable dye lasers and semiconductor diode lasers with high spectral resolution are often implemented. Photomultipliers with high quantum efficiency and high gain are commonly used as detectors.

Thin films or adsorbed overlayers on surfaces with thickness d contribute to the optical signal as d/λ with λ the wavelength of the exciting radiation. Hence optical probes are much more sensitive to surface coverage as compared with those in the IR region.[1] For example, changes in reflectivity below 10^{-4} are measurable, allowing one to detect a single atomic layer at optical wavelengths, $\lambda < 1$ µm.

In this chapter we shall consider linear optical techniques which are used for the investigation of surfaces and interfaces. They involve a linear susceptibility of the sample, $\chi^{(1)}$. Nonlinear optical techniques will be discussed in Chapter 6.

5.1
Spectroscopic Ellipsometry (SE)

In Section 3.1.1 we have discussed the phase changes on reflection. For the two-phase model of a crystal–vacuum interface which does not take into account any transition layer, the complex bulk dielectric function of the crystal

---

1) An exception is the case where the IR radiation is resonant to excitations in the overlayer.

Optics and Spectroscopy at Surfaces and Interfaces. Vladimir G. Bordo and Horst-Günter Rubahn
Copyright © 2005 WILEY-VCH Verlag GmbH & Co. KGaA, Weinheim
ISBN: 3-527-40560-7

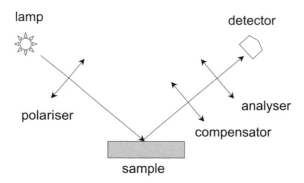

Fig. 5.1 Schematic sketch of an ellipsometer used in the optical region.

can be expressed in terms of the measured parameters as

$$\epsilon = \sin^2 \theta_i \left[1 + \tan^2 \theta_i \left(\frac{1 - \rho}{1 + \rho} \right)^2 \right]$$

(5.1)

where θ_i is the angle of incidence and ρ is the ratio of the reflection coefficient in p-polarization to that in s-polarization. This is the basic equation in spectroscopic ellipsometry (SE) which is widely used for studies of crystal surfaces and thin films in the optical region providing information on their composition and structure.

Any real surface contains a layer whose optical properties differ from those in the bulk crystal. That may be a thin film on the surface, in particular an oxide film, contamination, relaxed or reconstructed layer, or surface roughness. Therefore with the help of Eq. (5.1) an effective dielectric function, $\langle \epsilon \rangle$, is determined, which corresponds to an average over the region penetrated by the incident light. In order to extract the optical properties of a transition layer, the substrate contribution to $\langle \epsilon \rangle$ must be evaluated. This is usually performed by applying a three-phase model (see Section 3.1.3). Then the ellipsometric ratio, ρ, can be written using Eq. (3.40). The complex dielectric function (its real and imaginary parts) and the thickness of the transition layer (phase 2) are considered as the three unknown parameters. However, the measurements of the complex quantity ρ provide only two equations for them. To obtain the third one, it is necessary to invoke additional, physically reasonable restrictions.

In general, an ellipsometer used in the optical range (Fig. 5.1) resembles that of a FTIR ellipsometer (see Section 4.1). However, in this case the light sources are mainly Xe high-pressure arc-lamps and tungsten halogen lamps. The compensator is a phase-shifting element which is necessary to minimize the intensity at the detector in Null ellipsometry. Photometric ellipsometers, which are widely used in SE, measure the intensity of light reflected from a sample while modulating the polarization of the incident or reflected light. If

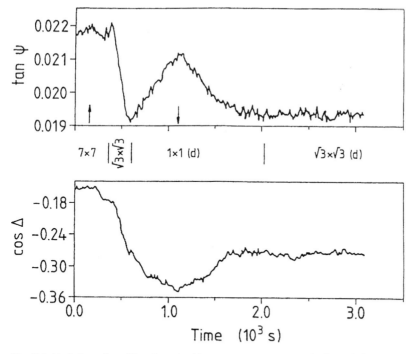

Fig. 5.2 Variation of tan Ψ and cos Δ with time during Ga adsorption on Si(111) at 500 °C. The arrows show where the shutter was opened and closed. Crystallographic structures are indicated along the time axis. Reprinted from (Andrieu and d'Avitaya 1991), Copyright 1991, with permission from Elsevier.

an analyzer is rotating at a constant frequency, the intensity at the detector can be expressed in terms of its transmission axis angle, θ_A, as

$$I(\theta_A) = I_0(1 + a_2 \cos 2\theta_A + b_2 \sin 2\theta_A) \qquad (5.2)$$

where the coefficients a_2 and b_2 are obtained by a Fourier transform of the intensity and are related to ρ and the polarizer angle, θ_P. An important advantage of photometric SE is that these Fourier coefficients are determined by relative intensities and hence are not sensitive to variations in the absolute intensity of the light source. Additional data, specific for anisotropic samples, can be obtained with variable angle SE (VASE) where both the angle of incidence and the wavelength are scanned.

SE is a sensitive optical tool for studying various surface properties and processes which occur on surfaces. It can be used to monitor *in situ* sample preparation, surface temperature, adsorption and growth. In particular, SE allows one to characterize thin metallic layers on semiconductor surfaces from the submonolayer to the nanometer range. This is relevant to metal contacts and Schottky barriers in various devices. Figure 5.2 shows the variation of the ellipsometric parameters with time during gallium (Ga) adsorption on the

Si(111) surface. The reflected intensity was measured with fixed wavelength at the Brewster angle where the bulk reflectivity of p-polarized light is minimum. The study was combined with reflection high-energy electron diffraction (RHEED) measurements to identify the crystallographic structure of the adsorbed layer. At small coverages, the adsorbate structure is (7×7) which is accompanied by a ($\sqrt{3} \times \sqrt{3}$) structure at the 1/10 monolayer (ML) Ga. At 2/3 ML a (1×1) structure appears. After the shutter is closed, gallium desorbs, leading to the reappearance of the ($\sqrt{3} \times \sqrt{3}$) pattern. These processes are manifested in different slopes of the ellipsometric parameters during the Ga adsorption/desorption process.

5.2
Reflection Difference Techniques

Whereas SE measures the ratio of reflection coefficients for different polarizations, various reflection difference techniques probe relative differences in reflectivity. Among these techniques one distinguishes *surface differential reflectivity* (SDR), *surface photoabsorption* (SPA) and *reflection anisotropy spectroscopy* (RAS).

5.2.1
Surface Differential Reflectivity (SDR)

In SDR one obtains the relative difference in reflectivities measured at near-normal incidence in p-polarization given by

$$\frac{\Delta R_{pp}}{R_{pp}} = \frac{R_{pp}^{clean} - R_{pp}^{ads}}{R_{pp}^{ads}} \tag{5.3}$$

where R_{pp}^{clean} and R_{pp}^{ads} are the reflectivities of the clean and adsorbate-covered surfaces, respectively, and the double subscript pp means that both the incident light and the reflected light are p-polarized.[2] One usually chooses the plane of incidence parallel to a principal axis of the sample surface. The adsorbate plays the role of a transition layer, and for an isotropic sample the quantity defined by Eq. (5.3) is equivalent to that in Eq. (3.49). If the sample is a semiconductor and the excitation energy is below the allowed energy gap, then Im(ϵ_3)=0 and the SDR signal is proportional to Im(ϵ_2), i.e., it is determined by the absorption in the overlayer.

The basic components of the experimental setup used for SDR are the same as those for SE (Section 5.1). Relative changes in reflectivity by as much as 10^{-3}

---

2) Generally, the polarization of light reflected from an anisotropic
   sample differs from that of the incident light (see Section 3.1.3).

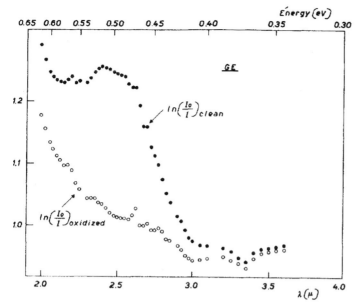

Fig. 5.3 Logarithm of the ratio of the incident intensity to the reflected one, I_0/I, as a function of wavelength in Ge(111)2×1 for the clean surface and for the same surface after oxidation. Reprinted with permission from (Chiarotti et al. 1971). Copyright 1971, American Physical Society.

are measured with the main experimental limitations dictated by the intensity fluctuations of the light sources. The sensitivity can be increased down to 10^{-4} relative reflectivity changes by using phase-sensitive detection methods and multiple total internal reflection geometries.

Figure 5.3 shows the ratio of the incident to the reflected light intensity at the clean Ge(111)2×1 surface and at the same surface after oxidation measured for photon energies below the allowed energy gap. A clean Ge surface, obtained by a cleavage in ultra-high vacuum, possesses dangling bonds which become saturated on oxygen adsorption. Therefore the difference between the two plots (Fig. 5.4) gives the surface absorption coefficient associated with intrinsic surface electronic states of a clean surface (Section 2.1.2). Their contribution has a maximum near 0.5 eV, indicating the transition energy between surface electronic states.

A special case of SDR which uses p-polarized light at an angle of incidence at or near the Brewster angle is known as the surface photoabsorption (SPA) technique. In a two-phase model the reflectivity in such a geometry is equal to zero. For a real interface, however, the reflected light intensity has a minimum at the Brewster angle but does not vanish (see Problem 3.3.). The value of this minimum reflected intensity is therefore very sensitive to the interface properties.

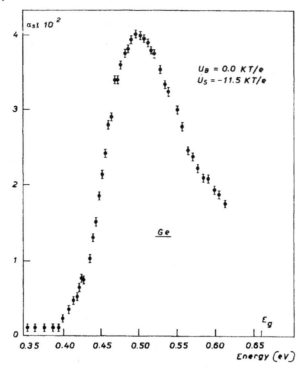

Fig. 5.4 Surface absorption constant versus photon energy calculated from the data shown in Fig. 5.3. Reprinted with permission from (Chiarotti et al. 1971). Copyright 1971, American Physical Society.

5.2.2
Reflection Anisotropy Spectroscopy (RAS)

Reflection anisotropy spectroscopy (RAS) probes the difference between the reflection coefficients measured at near-normal incidence for two mutually perpendicular polarizations. Let ψ be an azimuthal angle between the plane of incidence and one of the principal axes of the sample surface. Then the RAS signal normalized to the mean reflection coefficient, r, can be written as

$$\frac{\Delta r}{r} = 2 \frac{r_{pp}(\psi = 90°) - r_{pp}(\psi = 0°)}{r_{pp}(\psi = 90°) + r_{pp}(\psi = 0°)} \tag{5.4}$$

For an isotropic sample, $\Delta r = 0$, and therefore RAS measures the optical anisotropy. Another conclusion from here is that for an isotropic substrate only the overlayer contributes to the RAS signal. Such conditions are realized, for example, for the (001) surface of a cubic crystal, where one would measure the reflection coefficients for light polarized along the [110] and [$\bar{1}$10] directions (see Problem 5.3).

As with SDR, the basic components of the experimental setup for RAS measurements resemble those used in SE with the addition of a photoelastic modulator. The latter device advances and retards, sinusoidally, the phase of the reflected light. Then a Fourier analysis of the signal allows one to obtain the real and imaginary parts of the reflection anisotropy, $\Delta r / r$. Detection of changes in the RAS signal at a fixed photon energy takes less than 100 ms.

Being originally developed for the study of semiconductor surfaces, more recently RAS has been extended to probe the surface optical properties of metals and to monitor the growth of molecular assembly on metal surfaces (Weightman 2001). Self-assembled monolayers (SAM) on surfaces, especially those consisting of conjugated polymer molecules, are of considerable interest for applications in nanotechnology and molecular electronics. Figure 5.5 shows the evolution of the RAS signal obtained from a Cu(110) surface during deposition of 3-thiophene carboxylate (3TC) molecules. The changes observed if one uses photon energies of 2.1 eV are similar to those observed for the adsorption of other molecules on Cu(110) and thus are associated with the removal of intrinsic surface electronic states. As follows from electronic structure calculations, the RAS spectrum for photon energies of 4.25 eV is determined by transitions involving the Cu-carboxylate bonds. Notable changes in the signal at this energy correspond to the formation of an ordered c(4×8) structure with the plane of the thiophene ring parallel to the plane of the Cu-carboxylate bonds. With a further increase of coverage this structure changes to a p(2×1) symmetry where the thiophene ring is rotated away from the [110] direction. As one can see, this phase transition is not displayed in the RAS spectrum which means that there is no contribution to the RAS signal from the thiophene ring. A similar study of the adsorption of 9-anthracene carboxylic acid on Cu(110) has demonstrated, however, that RAS is capable of detecting the azimuthal orientation of a molecule which has a strong intramolecular transition in the spectroscopic range which is accessible to the RAS instrument.

5.3
Transmission Spectroscopy

If a substrate is transparent in the frequency range under consideration, the spectrum of an overlayer can also be measured in transmission (Fig. 5.6). In the optical region, this condition is fulfilled for dielectrics and therefore adsorption on dielectric surfaces can be studied in such a geometry. The sensitivity of the spectrum to adsorbate properties can be increased by passing light several times through similar surfaces.

Figure 5.7 shows the absorption spectrum of sodium adsorbed on a sapphire surface obtained in transmission through a pile of 17 polished sapphire

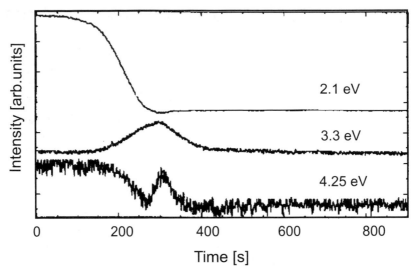

Fig. 5.5 Evolution of the RAS signal during increasing coverage by 3-thiophene carboxylic acid measured for three photon energies. Reprinted with permission from (Frederick et al. 1998). Copyright 1998, American Physical Society.

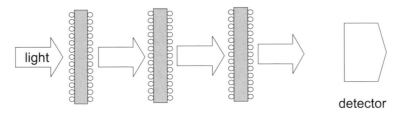

Fig. 5.6 Transmission geometry for measurements of an adsorbate absorption spectrum. To increase the sensitivity, a stack of samples can be used.

plates (Bonch-Bruevich et al. 1985). The two curves correspond to different surface temperatures. For comparison, the absorption spectrum of sodium vapor, containing both Na atoms and Na_2 molecules, is also reproduced. The broad absorption band of the adsorbate originates from transitions between vibrational states in the adsorption potential corresponding to the ground atomic state and those in the adsorption potential corresponding to the excited state (Fig. 5.8). Its broadening with the surface temperature is due to the population of higher vibrational sublevels in the ground potential well. The absence of vibrational structure in the spectrum is explained by a fast vibrational relaxation responsible for the broadening of vibronic transitions.

The measured dependencies of the surface absorption coefficient on both the vapor pressure and the surface temperature corresponded to the Lang-

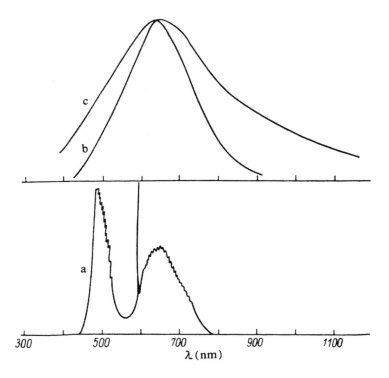

Fig. 5.7 (a) The absorption spectrum of sodium vapor, containing Na_2 molecules and Na atoms. The sharp line is the atomic transition. (b) The absorption spectrum of sodium adsorbed on a sapphire surface normalized to its maximum. $T = 470$ K. (c) The same as (b), but for $T = 670$ K. Reprinted with permission from (Bonch-Bruevich et al. 1985). Copyright 1986, American Institute of Physics.

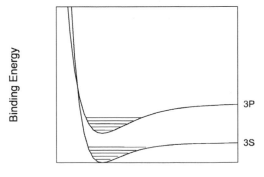

Fig. 5.8 Schematic representation of the energy levels of an adsorbed Na atom. The two potential wells correspond to adsorption of Na in the ground (3S) and excited (3P) electronic states. The sublevels in both potentials are given by the energies of the vibrational motion of an atom relative to the surface.

muir adsorption isotherm (see Section 2.2.5). This allowed the determination of the adsorption (binding) energy of Na adatoms which was found to be $E_b = 0.75\,\text{eV}$.

5.4
Photoluminescence Spectroscopy

Photoluminescence (PL) is the emission of light from a material following its illumination. The frequencies of both absorption and emission in luminescence are determined by the transitions between the electronic states, i.e., they correspond to the visible, or close to it, region of the spectrum. The rapidly decaying luminescence typical for atoms and molecules is usually called fluorescence. The lifetime, τ, of an excited electronic state can be represented as originating from two competitive channels, radiative versus nonradiative, as

$$\frac{1}{\tau} = \frac{1}{\tau_r} + \frac{1}{\tau_{nr}} \tag{5.5}$$

where τ_r and τ_{nr} are the lifetimes relative to radiative and nonradiative decay, respectively. The ratio between τ_r and τ_{nr} determines the luminescence efficiency. The increase of the nonradiative decay rate leads to a quenching of the luminescence. The applications of PL spectroscopy are specific for different surfaces and interfaces and we shall consider them separately.

5.4.1
Fluorescence Spectroscopy of Adsorbed Atoms and Molecules

The intensity of light emitted by lasers usually fluctuates in the percentage range. This limits the sensitivity of the absorption spectroscopy of adsorbates. An additional problem arises from the laser light scattered by the surface which has an effect on the signal. These disadvantages can be overcome if one observes spectrally shifted light emitted by atoms or molecules under investigation. For example, for an adsorbed atom such a situation occurs if the maxima of the absorption and emission bands do not coincide with each other. Another convenient way is to excite the adsorbate via two-photon absorption. Implementations of this excitation scheme will be considered in Section 6.5.

If fluorescence is excited by a laser pulse of a duration shorter than the lifetime of the excited state of an adsorbate, the subsequent decay of the emission intensity can be measured by time-resolved techniques. Then the slope of the logarithm of intensity versus time gives the total decay rate, $1/\tau$. The part related to nonradiative relaxation is determined by the energy transfer to the substrate. Thus such measurements allow one to study different de-excitation channels at a surface.

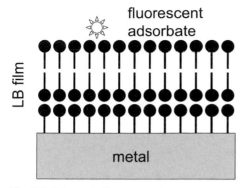

Fig. 5.9 Schematic illustration of measurements of an excited-state lifetime versus adsorbate–surface distance. A spacer layer consists of a number of Langmuir–Blodgett monolayers each of which has a well-defined thickness.

Atoms or molecules adsorbed on a metal surface do not fluoresce because of highly effective quenching via creation of electron–hole pairs in the substrate. However, their fluorescence can be observed if they are separated from a metal by a dielectric spacer layer. The Langmuir–Blodgett technique allows one to deposit a fatty acid layer of a well-controlled thickness[3] and thus provides an opportunity to study fluorescence of an adsorbate versus its distance from the surface (see Fig. 5.9).

Figure 5.10 shows a comparison between experimental data and theory for the measurements of the excited-state lifetime of fluorescent molecules near a silver surface. At large distances, τ tends to the radiative lifetime of an isolated molecule, $\tau_0 = 632\ \mu s$. When a molecule moves closer to the surface, oscillations of the lifetime occur, resulting from interference between the light emitted by a molecule and that reflected from the surface. When the molecule–surface distance decreases further ($d < 200\ \text{Å}$), the lifetime abruptly decreases. This indicates that an additional channel of de-excitation, which is related to the nonradiative energy transfer to the metal substrate, comes into play. In this case, the relaxation of the excited state is described as the decay of the molecular image and its rate varies as $1/d^3$ with the molecule–surface distance (see Section 2.2.4). At a rough metal surface, however, this behavior can be considerably modified (Leung and George 1989).

3) Each fatty acid layer has a thickness of 26.4 Å determined by the length of a single molecule.

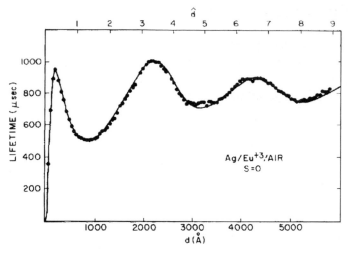

Fig. 5.10 The lifetime of dye molecules containing an Eu^{+3} ion (emission at 612 nm) at a silver surface as a function of the ion–surface distance. The experimental data are shown by dots, the solid line represents a fit. Reprinted with permission from (Chance et al. 1975a). Copyright 1975, American Institute of Physics.

5.4.2
Photoluminescence Spectroscopy at Semiconductor Surfaces and Interfaces

In semiconductors, PL originates from the radiative recombination of photoexcited electron–hole pairs. Their nonradiative lifetime is determined by both bulk and surface recombination. Therefore, the major problem in observing PL from semiconductor surfaces and interfaces is to minimize signals which arise from defects and impurities in the bulk. This can be achieved with an improvement in material quality by means of an epitaxial growth together with the fabrication of special structures, such as semiconductor heterostructures and quantum wells. The control of the epitaxial growth process allows one to vary the relative contribution of bulk and surface recombination independently of each other.

PL can be excited by a lamp. However, lasers provide higher brightness and coherence. Pulsed lasers can be used for time-resolved measurements. As phonons generally contribute to relaxation processes, the sample under investigation is usually mounted in a cryostat allowing one to vary the temperature from 4.2 K. For detection of emitted light below 1000 nm one uses photomultipliers or cooled CCD (charge-coupled-device) cameras, whereas at longer wavelengths liquid-nitrogen-cooled Ge detectors are used.

PL is an extremely sensitive probe of surface recombination properties. Being combined with a careful *in situ* control of surface reconstruction and chemical termination, it allows one to assess how these factors influence the surface

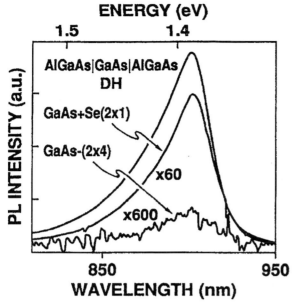

Fig. 5.11 Typical *in situ* photoluminescence spectra from GaAs incorporated in a double heterostructure compared with the spectra from a bare (2×4) and a Se-passivated (2×1) GaAs surface. Reprinted with permission from (Sandroff et al. 1991). Copyright 1991, American Institute of Physics.

electronic quality. Figure 5.11 compares typical PL spectra obtained *in situ* from both a bare and a Se-passivated GaAs surface with that from GaAs incorporated in a double heterostructure AlGaAs/GaAs/AlGaAs. Due to a high density of mid-gap surface states, PL intensities from exposed GaAs surfaces are 10^3–10^4 times lower than from an AlGaAs-capped surface. Therefore, the latter structure can be used as a test sample allowing one to estimate the quality of modified GaAs surfaces.

5.5
Raman Spectroscopy

In Raman spectroscopy, the spectrum of light inelastically scattered by a sample is registered. This process is called Raman scattering (RS) if the frequency of the incident light is in the visible region of the spectrum and the frequency shift observed in scattering corresponds to vibrational or rotational transitions of the material. Raman spectra are widely used for the identification of chemical compounds in the sample as well as for the study of vibrational dynamics. Similar to photoluminescence spectroscopy, Raman spectroscopy is not

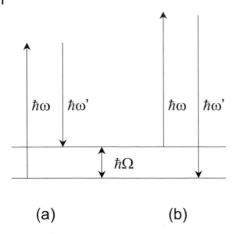

Fig. 5.12 Schematic diagram of transitions in Stokes (a) and anti-Stokes (b) Raman scattering.

an intrinsic surface-specific technique. However, in principle, surface properties can be extracted from Raman spectra when they are distinguishable from those in the bulk. RS has some peculiarities at metal surfaces and we shall consider that case separately.

5.5.1
Raman Scattering

The RS of light is accompanied by the creation or annihilation of an elementary excitation in the sample. It may be an optical phonon in a crystal or a vibration of an adsorbed molecule. As a result, the frequency of the scattered light, ω', is different from the frequency of the incident light, ω. As follows from energy conservation in a scattering process of incident photons,

$$\hbar\omega' = \hbar\omega \pm \hbar\Omega \tag{5.6}$$

where Ω is the frequency of the elementary excitation. The minus sign in Eq. (5.6) implies the creation of excitation (Stokes scattering), whereas the plus sign corresponds to annihilation (anti-Stokes scattering). These processes are illustrated in Fig. 5.12. The intensities of both lines are proportional to the populations of the initial states, therefore in thermal equilibrium the anti-Stokes line is much less intensive than the Stokes line.

RS originates from the coupling between electronic and vibrational motions in the material. This effect is described classically in terms of the linear susceptibility tensor, $\chi_{ij}^{(1)}$, which relates the sample polarization, **P**, and the electric field of the light wave, **E**, as

$$P_i(\omega') = \chi_{ij}^{(1)}(\omega, \omega')E_j(\omega) \tag{5.7}$$

The susceptibility is a function of the normal coordinates, Q_r, associated with the vibrational motion, and can be expanded in a Taylor series in them, i.e.,

$$\chi_{ij}^{(1)}(Q_r) = \chi_{ij}^{(1)}(0) + \sum_r \left(\frac{\partial \chi_{ij}^{(1)}}{\partial Q_r}\right)_0 Q_r + \dots \tag{5.8}$$

where the zero denotes the equilibrium nuclear configuration. The first term in the expansion (5.8) corresponds to elastic scattering of light, whereas the second one, being multiplied by $E_j(\omega)$, describes the frequency modulation of the induced polarization by vibrations of the nuclei.

In quantum theory, the intensity of Raman scattering from the initial ground state $|g\rangle$ to the final excited state $|e\rangle$ is directly proportional to the incident light intensity as well as to the squared modulus of the scattering tensor component

$$(\alpha_{ij})_{ge} = \frac{1}{\hbar}\sum_k \left[\frac{(\mu_j)_{gk}(\mu_i)_{ke}}{\omega_{kg} - \omega} + \frac{(\mu_i)_{gk}(\mu_j)_{ke}}{\omega_{ke} + \omega}\right] \tag{5.9}$$

where μ_i are the Cartesian components of the dipole moment operator. The summation runs over the intermediate excited states $|k\rangle$, ω_{mn} are the transition frequencies between the corresponding states. The symmetry properties of the sample require that some matrix elements of the tensor α_{ij} are equal to zero thus determining selection rules for RS. If the system under investigation is invariant relative to inversion of all coordinates, no transition appearing in the Raman spectrum can appear in the infrared absorption spectrum and vice-versa.

RS has a small cross-section; typically, one Stokes photon is produced per 10^{10} incident photons. This fact limits applications of Raman spectroscopy for surface and interface analysis. However, the RS efficiency can be considerably increased if the incident light frequency, ω, is close to a transition frequency ω_{kg}. Then the corresponding term in the tensor (5.9) is enhanced, leading to resonance Raman scattering. In order to obtain a detectable Raman signal one has to apply a powerful light source. Therefore, in contrast to other linear optical techniques, the use of lasers is essential in Raman spectroscopy.

Raman spectroscopy can be successfully applied for studying overlayers that form an interface with the substrate without interdiffusion between overlayer and substrate. Electronic properties, lattice dynamics and its influence on the substrate, can be investigated. Raman spectroscopy also allows one to identify new phases, a few atomic layers thick, at the interface with a reactive overlayer.

5.5.2
Surface-enhanced Raman Scattering (SERS)

In 1974 it was discovered that RS from pyridine molecules adsorbed on a rough silver surface has an enormously large cross-section. This phenomenon received the name *surface-enhanced Raman scattering* (SERS). The initial approach was to roughen the sample surface to increase the surface area and, hence, the number of adsorbed molecules available for study. However, a few years later it was recognized that the large Raman intensities observed, which exceed the usual values by 5–8 orders of magnitude, cannot be attributed to the larger number of scatterers. This finding stimulated a wealth of investigations, both experimental and theoretical.

SERS has been observed for a large variety of molecules adsorbed on the surfaces of relatively few metals; mainly on silver, copper and gold, for different morphologies and physical environments. The largest enhancements occur for surface roughnesses in the nanoscale range (10–100 nm). Various features of SERS differ from those typical for ordinary RS. The intensities of the Raman bands generally decrease with increasing vibrational frequency. Since the selection rules are less tight, modes normally forbidden in Raman spectra appear in SERS. The spectra are almost completely depolarized, even for molecules adsorbed on atomically smooth surfaces. The enhancement extends to long distances, up to tens of nanometers, from the substrate. Possible mechanisms responsible for these features can be broadly classified as *electromagnetic* and *chemical*.

We have seen in Section 3.3 that light striking a rough surface excites surface polaritons (surface plasmon polaritons in the case of a metal surface). Surface protrusions and hollows support localized surface polaritons. The SP electromagnetic field is strengthened in comparison with the incident light field. The intensity enhancement factor, $\eta(\omega)$, depends on the surface geometry. This provides a basis for the explanation of SERS from the viewpoint of electromagnetic enhancement. A rough surface can be modelled in different ways: as isolated or interacting spheres and ellipsoids, or as a randomly rough surface. In all cases, the surface roughness not only amplifies the incident light field, $E_i(\omega)$, but also the Raman scattered field, $E_s(\omega')$ (see Fig. 5.13). The overall Raman intensity enhancement factor is then determined by the product $\eta(\omega)\eta(\omega')$.

This model also explains qualitatively some of the other experimental observations. A limited number of substrates active in SERS implies that the frequencies of localized SPs for them are in the visible spectral range commonly used for Raman spectroscopy. In addition, the imaginary part of the dielectric function for those metals is very small at the SP frequency, ensuring small losses in the substrate. As the SERS enhancement is proportional to $\eta(\omega)\eta(\omega')$, it acquires large values only if the resonance condition with lo-

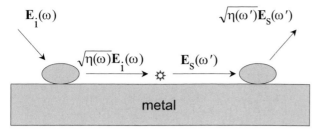

Fig. 5.13 Scheme of electromagnetic field enhancement at a rough metal surface.

calized SPs is satisfied for both frequencies, ω and ω', i.e., the Raman shift is relatively small. The long-range character of the effect originates from its electromagnetic nature, whereas the depolarization of SERS spectra can be attributed to different local environments of adsorbed molecules, as well as to a variety of their orientations.

However, some experimental data suggest that there is another enhancement mechanism independent of the one considered above. Electromagnetic enhancement should be nonselective with respect to different molecules adsorbed on the same substrate. This is not the case for CO and N_2 molecules; their SERS intensities differ by a factor of 200 under the same experimental conditions. The polarizabilities of these molecules are nearly identical and no orientational effect can explain such a large intensity difference. Another evidence supporting a chemical mechanism is a broad resonance observed when one scans the incident light frequency. These observations can be explained by a resonance Raman scattering via electronic states which are formed in chemisorption. The chemical bond can be described in terms of the HOMO and LUMO (see Section 2.2.2) symmetrically disposed in energy with respect to the Fermi level of the metal. Then charge-transfer excitations can occur at about half the energy of the intrinsic molecular transition between the HOMO and LUMO. For molecules studied by SERS the frequency of the intramolecular transition is typically in the near ultraviolet and, hence, the charge-transfer transitions are in the visible region of the spectrum.

Generally, it is very difficult to identify the chemical enhancement due to its small contribution as compared with the contribution from the electromagnetic enhancement. Figure 5.14 shows as an exception the Raman spectra of pyromellitic dianhydride (PMDA) adsorbed on Cu(111) excited at 647 and 725 nm. The lower spectrum is clearly enhanced as compared with the upper one. Figure 5.15 represents the electronic absorption spectrum of the same system obtained with electron energy loss spectroscopy (EELS). The intense narrow peak at 1.9 eV (the transition wavelength 653 nm) appears only for copper covered with a monolayer of PMDA, whereas the intrinsic intramolecular ex-

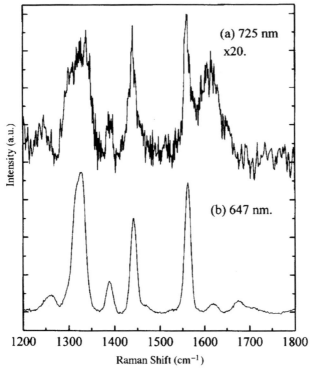

Fig. 5.14 Raman spectra of pyromellitic dianhydride adsorbed on Cu(111) excited at 725 nm (a) and 647 nm (b) (Campion and Kambhampati 1998). Reproduced with permission of The Royal Society of Chemistry.

citations of PMDA occur in the ultraviolet region. These results demonstrate that a new low-energy electronically excited state created in chemisorption of PMDA on copper is responsible for the enhancement of the Raman spectrum.

5.6
Surface Plasmon Polariton Spectroscopy

In Section 4.3 we have considered surface polariton spectroscopy in the infrared region. As distinct from that case, SPs in the optical spectral range exist mainly on metal surfaces, i.e., as surface plasmon polaritons (SPP).[4] The SPP dispersion curve extends from the far-infrared up to the far-ultraviolet region and thus it can be excited at any frequency in this range. This allows one to use SPPs as an optical probe of various overlayers on metal surfaces and in-

---

4) Another example is surface exciton polaritons on semiconductor surfaces.

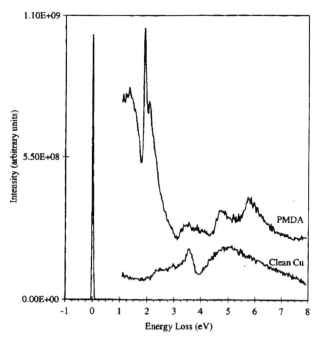

Fig. 5.15 Electron energy loss spectrum of pyromellitic dianhydride adsorbed on Cu(111); the spectrum of clean Cu(111) is also shown for comparison (Campion and Kambhampati 1998). Reproduced with permission of The Royal Society of Chemistry.

terfaces between metals and another media. The SP propagation lengths in the optical region are much less than in the infrared. Therefore, in SPP optical spectroscopy one usually measures the reflection coefficient in attenuated total reflection (ATR) geometry rather than the SPP propagation length.

5.6.1
Spectroscopy at Metal Surfaces and Interfaces

Any overlayer or a film deposited onto a metal surface modifies the SPP dispersion relation determined for a clean surface. If its thickness, d, is much less than the SPP wavelength, λ, the corresponding change in the SPP wave vector modulus can be written as (Abelès and Lopez-Rios 1982)

$$\Delta q = -\frac{4\pi^2 d}{\lambda^2}\frac{(-\epsilon_1\epsilon_2)^{3/2}(\epsilon_f - \epsilon_2)(\epsilon_1 - \epsilon_f)}{\epsilon_f(\epsilon_1 - \epsilon_2)(\epsilon_1 + \epsilon_2)^2} \tag{5.10}$$

where ϵ_1, ϵ_2 and ϵ_f are the dielectric functions of the surrounding medium, metal and film, respectively, which are complex quantities in a general case. In ATR experiments, $\mathrm{Re}(\Delta q)$ determines the shift of the ATR dip due to the presence of the film, whereas $\mathrm{Im}(\Delta q)$ is related to its broadening. These quan-

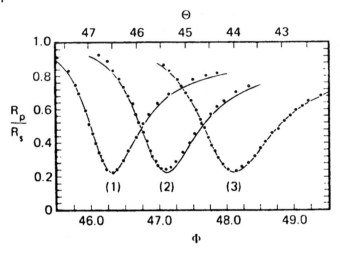

Fig. 5.16 ATR curves obtained in Kretschmann geometry at $\lambda = 6328$ Å at a clean Au surface (1) and at the same surface covered with two (2) and four (3) layers of cadmium arachidate. Reprinted from (Gordon and Swalen 1977). Copyright 1977, with permission from Elsevier.

tities can be expressed in terms of the shift of the angular position of the ATR minimum, $\mathrm{Re}(\Delta\theta)$, and its broadening, $\mathrm{Im}(\Delta\theta)$, through the equation

$$\Delta q = \frac{2\pi}{\lambda} n_p \cos\theta \Delta\theta \tag{5.11}$$

with n_p the index of refraction of the prism. The change of the dip depth is also connected with $\mathrm{Im}(\Delta q)$ and, hence, with absorption in the film.

Figure 5.16 shows the results of ATR measurements, normalized to almost constant reflection coefficient in s-polarization, at a clean Au surface as well as at the same surface covered with different numbers of cadmium arachidate layers. The value of the reflection coefficient in the minimum is not influenced essentially by the overlayer which implies the realness of the dielectric function of the organic film.

Figure 5.17 represents a qualitatively different situation. Here, the Ag substrate is covered with Pd films which are characterized by a large $\mathrm{Im}(\epsilon_f)$. With increasing film thickness, the ATR dip broadens markedly and its depth is reduced.

SPP spectroscopy also allows one to study interfaces between a metal and another solid. If a metal is in contact with a transparent medium with a dielectric function ϵ_1, SPP can be excited in the ATR method using a prism with the index of refraction n_p such that $n_p^2 > \epsilon_1$. In stratified media, both surfaces of a metal layer can be studied by independent excitation of SPPs at each of them. Figure 5.18b schematically represents the SPP dispersion curves for the system shown in Fig. 5.18a. If the metal film is thick enough, the SPPs on

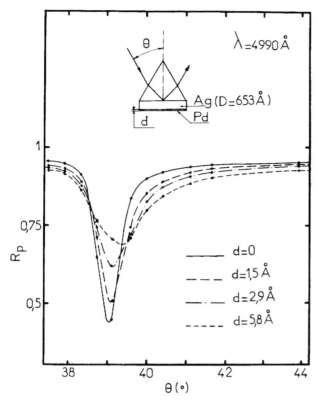

Fig. 5.17 ATR curves obtained in Kretschmann geometry at $\lambda = 4990$ Å from Ag covered with Pd layers of different thickness. Reprinted with permission from (Abelès and Lopez-Rios 1982).

different surfaces do not interact with each other. Then SPP can be excited at each surface when the angle of incidence is varied at a fixed frequency, or, alternatively, when the frequency is varied at a fixed angle of incidence. This results in two ATR minima as shown in Fig. 5.18c.

SPPs can be excited at an interface between two metals. In the Drude model their dielectric functions, ϵ_1 and ϵ_2, are determined by their plasma frequencies, ω_{p1} and ω_{p2}. If $\omega_{p1} > \omega_{p2}$, then in the frequency range $\omega_{p2} < \omega < \omega_{p1}$ one has $\epsilon_2 > 0$ and $\epsilon_1 < 0$, i.e., the condition of the existence of SPP is fulfilled. It is worthwhile to note that, at such frequencies, $\epsilon_2 < 1$ and, hence, total internal reflection, necessary for SPP excitation, is realized at an interface between metal 2 and a vacuum without a prism.

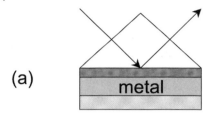

(a)

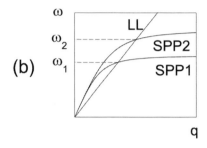

(b)

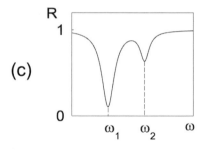

(c)

Fig. 5.18 (a) Schematic diagram of the ATR method in a stratified system. The dielectric layers are transparent and the metal film is so thin that the incident light can penetrate through it. (b) Two SPP dispersion curves cor-responding to different surfaces of a metal film. LL is the light line described by the equation $\omega = ck_x/(n_p \sin \theta_i)$. (c) Qualitative behavior of the reflection coefficient.

5.6.2
Spectroscopy at Electrochemical Interfaces

The excitation of a SPP is also a valuable tool for the study of an electrochemical interface. It can be used *in situ* in an electrochemical cell and can provide information on the structure and electronic properties of the electrode–electrolyte interface in the region of transparency of the solution (~ 1 eV to ~ 6 eV). This method is especially useful for the investigation of adsorption isotherms because the chemical potential of an adsorbate and, therefore,

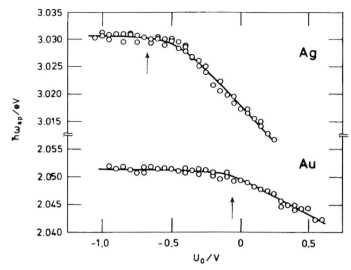

Fig. 5.19 Surface plasmon polariton excitation energy for polycrystalline Ag and Au electrodes in 0.5 M NaClO$_4$ as a function of the electrode potential. The potentials of zero charge are marked by arrows. Reprinted from (Kötz et al. 1977), Copyright 1977, with permission from Elsevier.

the surface coverage can be reversibly varied over a wide range by means of changing the electric potential of the electrode.

Of particular interest is the question whether the electrode potential markedly influences the SPP excitation. Figure 5.19 shows the SPP excitation energy, $\hbar\omega_{sp}$ for Ag and Au electrodes as a function of the electrode potential. This energy is found from the ATR minimum by scanning the wavelength of the incident radiation. This dependence is remarkably distinct for different charging of the electrode. When the electrode is charged negatively (i.e., its potential is below the PZC) the SPP excitation energy does not depend on the potential, whereas when it is charged positively (the potential is above the PZC) $\hbar\omega_{sp}$ decreases linearly with the potential.

These features can be qualitatively explained on the basis of a jellium model for the electron distribution at the electrode–electrolyte interface (Fig. 5.20). The electron density varies smoothly at the interface, from its value in the bulk metal, to zero in the electrolyte. A change in the electrode potential shifts the tail of the electron density inwards (for positive potential) or outwards (for negative potential) relative to the positive background of the atomic cores. In the former case, the surface region of the metal is depleted of electrons, leading to a lowering of the plasma frequency that is "felt" by the SPP. In the latter case, the electron density near the metal surface changes only slightly.

The potential of SPP spectroscopy for the monitoring of electrochemical reactions is illustrated in Figure 5.21. The figure shows the change in the SPP

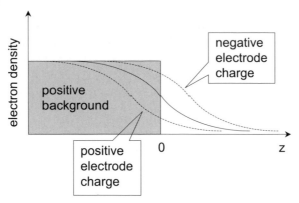

Fig. 5.20 The distribution of electron density at the electrode–electrolyte interface for different electrode potentials.

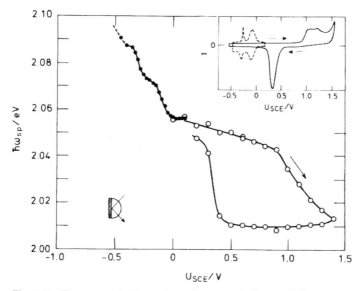

Fig. 5.21 The change in the surface plasmon polariton excitation energy during electrochemical reactions on Au in 0.5 M NaClO$_4$. The open and filled circles correspond to oxide formation and Pb monolayer formation, respectively. The inset is a simultaneously obtained voltammogram. Reprinted with permission from (Kolb 1982).

excitation energy during electrochemical reactions at the gold–electrolyte interface measured in Kretschmann geometry at fixed angle of incidence. The cyclic voltammogram shown in the inset allows one to follow the reaction cycles. One reaction is the formation of a surface oxide on gold by anodic decomposition of water. The oxide starts to form at +0.9 V and is reduced on the reversed potential scan near +0.3 V. This redox cycle is clearly seen in the variation of $\hbar\omega_{sp}$ with the potential. Another reaction occurred when

2×10^{-4} M $Pb(NO_3)_2$ were added to the solution and one monolayer of lead was deposited during the potential scan from 0 to -0.5 V. As seen from the voltammogram in the inset, between 0 and -0.5 V this reaction has two different steps which correspond to different adsorption energies. This feature is also reflected in the dependence of the SPP excitation energy on the electrode potential.

5.7
Electrochemical Optical Spectroscopy

The application of reflection spectroscopy to an electrode–electrolyte interface combines the advantages of electrochemical and optical techniques. One can change the equilibrium state of the electrode surface by varying the electrode potential and carry out *in situ* measurements with high accuracy. Based on this a variety of methods, which are generally called *electrochemical optical spectroscopy*, investigate the dependence of different parameters characterizing the light reflected from a metal electrode immersed in an electrolyte on the applied voltage, ϕ (Brodsky et al. 1985). Usually ϕ is modulated periodically in time:

$$\phi(t) = \phi_0 + \Delta\phi \sin \Omega t \qquad (5.12)$$

and the component which varies with the frequency Ω in the reflected signal is selected. If the modulation amplitude, $\Delta\phi$, is small, the reflection coefficient can be written as

$$r = r_0 + \left(\frac{dr}{d\phi}\right)_{\phi=\phi_0} \Delta\phi \sin(\Omega t + \delta') \qquad (5.13)$$

where $r_0 = r(\phi_0)$ and the phase shift δ' is nonzero when the potential variation is accompanied by slow processes (on a time scale $\sim \Omega^{-1}$) which alter the optical properties of the system. Modulation techniques include *electroreflectance* in which one measures the quantity $(1/R)(dR/d\phi)$ for different polarizations, *modulated ellipsometry*, where the derivatives of the ellipsometric angles, $d\Delta/d\phi$ and $d\Psi/d\phi$, are obtained, and the method of *electromodulation of the reflection phase* which provides the derivatives of the phase changes on reflection, $d\delta_s/d\phi$ and $d\delta_p/d\phi$. As an example, Fig. 5.22 shows the reflection spectrum obtained by the electroreflectance method from silver samples immersed in KCl-water electrolyte. The applied voltage of about 2 V peak-to-peak amplitude was modulated with a frequency of $\Omega = 35$ Hz. Two distinctive features in the electroreflectance spectrum are observed: a negative peak at the steep plasma edge and a narrow peak in the region where the metal has low reflectivity.

Information about the electrode–electrolyte interface can be extracted from electrochemical spectra if one establishes a correlation between its optical

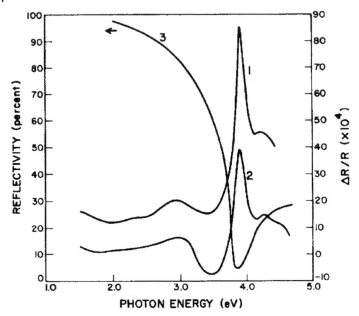

Fig. 5.22 Electroreflectance of silver for a high-purity polished sample (1) and high-purity polished commercial sheet silver (2). Curve (3) shows the reflectivity of sample (1) measured in air. Reprinted with permission from (Feinleib 1966). Copyright 1966, American Physical Society.

properties and observable quantities. In the linear regime of the light–interface interaction, the total optical signal is determined by the sum of the contributions from the different regions. The most informative is the essentially nonlocal contribution of the microscopic compact layer (Section 2.3.1). It can be expressed in terms of the nonlocal dielectric tensor of this region. In addition, the change in the potential decrease at the electrode–electrolyte interface is accompanied by a change in the ion concentration distribution in the diffuse layer of the solution and thus also by a change in the dielectric function of this region.

Another important point of data analysis is to find a relation between the dielectric properties of the electrode–electrolyte interface and its microscopic characteristics. A rigorous treatment of this problem is rather complex and involves large-scale computer calculations. An alternative method is to use semi-phenomenological models which relate the behavior of the dielectric tensor components to different features of the electron spectrum of the system. The mechanisms responsible for modulated electroreflectance can be classified as those arising from the modulation of the electron density in the selvedge region of the electrode (plasma electroreflectance), and from those which are due to a modulation of both interband and intraband optical transitions. A careful analysis of the electrochemical optical spectra based on theoretical models allows one also to extract information on the electron structure of particles adsorbed at the electrode, their vibrations, intermolecular interactions and charge transfer at the interface, as well as on the kinetics of adsorption.

Further Reading

V.M. Agranovich, D.L. Mills (Eds.), *Surface Polaritons*, North-Holland Publishing Company, Amsterdam, 1982.

R.M. Assam, N.M. Bashara, *Ellipsometry and Polarised Light*, North-Holland Publishing Company, Amsterdam, 1997.

A.M. Brodsky, L.I. Daikhin, M.I. Urbakh, Interpretation of data on electrochemical optical spectroscopy of metals, *J. Electroanal. Chem.* **1984**, *171*, 1.

A. Campion and P. Kambhampati, Surface-enhanced Raman scattering, *Chem. Soc. Rev.* **1998**, *27*, 241.

J.F. McGilp, Optical characterization of semiconductor surfaces and interfaces, *Prog. Surf. Sci.* **1995**, *49*, 1.

J.F. McGilp, D. Weaire, C.H. Patterson (Eds.), *Epioptics – Linear and Nonlinear Optical Spectroscopy of Surfaces and Interfaces*, Springer-Verlag, Berlin, 1995.

Problems

Problem 5.1. Estimate the phase change of a p-polarized light beam reflected from a Si(111) surface during Ga adsorption (see Fig. 5.2 and the corresponding discussion).

Problem 5.2. A thin film of thickness $d \ll \lambda$ is deposited onto a semiconductor surface. A light beam strikes the film from the vacuum side perpendicular to its surface. In the framework of the three-phase model derive an expression for the surface differential reflectivity at energies below the forbidden energy gap in terms of the film absorption coefficient.

Problem 5.3. Using Eq. (3.51), derive an expression for the RAS signal from an anisotropic layer deposited onto a surface of an isotropic crystal.

Problem 5.4. In which states, $3S$ or $3P$, is the adsorption energy of Na atoms adsorbed on a sapphire surface larger? Use the absorption spectra of both sodium vapor and sodium atoms adsorbed on a sapphire surface (Fig. 5.7).

Problem 5.5. Give arguments showing that the oscillations of the excited state lifetime with distance, given in Fig. 5.10, originate in the interference effect. Use $n = 1.5$ as the refractive index of the Langmuir-Blodgett film.

Problem 5.6. Estimate the maximum SERS intensity enhancement factor for $\lambda = 820$ nm at a rough silver surface based on the simulation of surface roughness by: (a) small spheres; (b) small spheroids with the depolarization factor $A = 0.1$. The index of refraction of silver at the given wavelength is $\tilde{n} = 0.04 + i \cdot 5.73$. For the estimate, neglect the variation of the dielectric function over the vibrational frequency shift.

6

Nonlinear Optical Techniques at Surfaces and Interfaces

6.1
Nonlinear Optical Response

Classification of nonlinear optical effects

So far we have considered the optical properties of surfaces and interfaces under the assumption that the amplitude of the incident light wave, $\mathbf{E}_i$, is negligible in comparison with the intra-atomic field strengths and hence the electric displacement vector is a linear function of $\mathbf{E}_i$ (see Eq. (3.63)).[1] For larger amplitudes the polarization which is induced by the light wave can be quite generally written in the form of a power series

$$\mathbf{P}(\mathbf{r}, t) = \mathbf{P}^{(1)}(\mathbf{r}, t) + \mathbf{P}^{(2)}(\mathbf{r}, t) + \mathbf{P}^{(3)}(\mathbf{r}, t) + \dots \qquad (6.1)$$

where the nth term is proportional to the nth power of the incident wave amplitude. The first term in this expansion refers to the linear optical response discussed in the preceding sections, whereas the subsequent terms describe various nonlinear optical phenomena.

The quantities $\mathbf{P}^{(n)}$ entering Eq. (6.1) are determined as

$$\mathbf{P}^{(1)}(\mathbf{r}, t) = \int \int_{-\infty}^{\infty} \chi^{(1)}(\mathbf{r}, \mathbf{r}_1; t - t_1) : \mathbf{E}_i(\mathbf{r}_1, t_1) d\mathbf{r}_1 dt_1 \qquad (6.2)$$

$$\mathbf{P}^{(2)}(\mathbf{r}, t) = \int \int \int_{-\infty}^{\infty} \int_{-\infty}^{\infty} \chi^{(2)}(\mathbf{r}, \mathbf{r}_1, \mathbf{r}_2; t - t_1, t - t_2) : \mathbf{E}_i(\mathbf{r}_1, t_1)$$
$$\times \mathbf{E}_i(\mathbf{r}_2, t_2) d\mathbf{r}_1 d\mathbf{r}_2 dt_1 dt_2 \qquad (6.3)$$

---

1) The typical intra-atomic field strengths are of the order 10^8–10^9 V/cm.

Optics and Spectroscopy at Surfaces and Interfaces. Vladimir G. Bordo and Horst-Günter Rubahn
Copyright © 2005 WILEY-VCH Verlag GmbH & Co. KGaA, Weinheim
ISBN: 3-527-40560-7

and so on, where $\chi^{(n)}$ are the nth order nonlinear susceptibility tensors. Due to the translational invariance along the surface or interface, the quantities $\chi^{(n)}$ depend on the differences between the vector components parallel to it, $\mathbf{r}_\| - \mathbf{r}_{1\|}, \mathbf{r}_\| - \mathbf{r}_{2\|}, \ldots$

If a group of monochromatic plane waves strikes the interface, the electric field $\mathbf{E}_i$ must be replaced by the sum[2]

$$
\begin{aligned}
\mathbf{E}(\mathbf{r}, t) &= \sum_i \mathbf{E}_{i0} \exp[i(\mathbf{k}_i \cdot \mathbf{r} - \omega_i t)] \\
&\equiv \sum_i \mathbf{E}(z; \mathbf{k}_{i\|}, \omega_i) \exp[i(\mathbf{k}_{i\|} \cdot \mathbf{r}_\| - \omega_i t)]
\end{aligned}
\tag{6.4}
$$

Then, introducing the Fourier-transformed quantities by means of the equation

$$
\mathbf{P}^{(n)}(\mathbf{r}, t) = \int \int_{-\infty}^{\infty} \mathbf{P}^{(n)}(z; \mathbf{k}_\|, \omega) \exp[i(\mathbf{k}_\| \cdot \mathbf{r}_\| - \omega t)] d\mathbf{k}_\| d\omega
\tag{6.5}
$$

we arrive at the expansion (6.1)[3]

$$
\mathbf{P}(\mathbf{k}_\|, \omega) = \mathbf{P}^{(1)}(\mathbf{k}_\|, \omega) + \mathbf{P}^{(2)}(\mathbf{k}_\|, \omega) + \mathbf{P}^{(3)}(\mathbf{k}_\|, \omega) + \ldots
\tag{6.6}
$$

where

$$
\mathbf{P}^{(1)}(\mathbf{k}_\|, \omega) = \sum_i \chi^{(1)}(\mathbf{k}_{i\|}, \omega_i) : \mathbf{E}(\mathbf{k}_{i\|}, \omega_i)\delta(\mathbf{k}_\| - \mathbf{k}_{i\|})\delta(\omega - \omega_i)
\tag{6.7}
$$

$$
\begin{aligned}
\mathbf{P}^{(2)}(\mathbf{k}_\|, \omega) &= \sum_{ij} \chi^{(2)}(\mathbf{k}_{i\|}, \omega_i; \mathbf{k}_{j\|}, \omega_j) : \mathbf{E}(\mathbf{k}_{i\|}, \omega_i)\mathbf{E}(\mathbf{k}_{j\|}, \omega_j) \\
&\quad \times \delta(\mathbf{k}_\| - \mathbf{k}_{i\|} - \mathbf{k}_{j\|})\delta(\omega - \omega_i - \omega_j)
\end{aligned}
\tag{6.8}
$$

and so forth, and $\chi^{(n)}$ are the Fourier transforms of the corresponding susceptibility tensors.[4]

The δ-functions with respect to frequency in the expressions for the nonlinear polarization terms represent energy conservation in the relevant nonlinear optical process. For example, if two electromagnetic waves with the frequencies ω_1 and ω_2 strike the interface, the second-order term, $\mathbf{P}^{(2)}$, describes the process of *sum-frequency generation* where a new electromagnetic wave with frequency $\omega = \omega_1 + \omega_2$ is generated. Not very different from this process is *difference-frequency generation* which leads to the generation of a wave with frequency $\omega = \omega_1 - \omega_2$. The corresponding term arises if one replaces the

---

2) The observable field is described by the real part of $\mathbf{E}(\mathbf{r}, t)$.
3) To simplify the notation in what follows we omit the argument z.
4) Strictly speaking, the quantities $\chi^{(n)}$ differ from the Fourier transforms by a factor of $(2\pi)^{3n}$.

amplitude $\mathbf{E}(\mathbf{k}_{2\parallel}, \omega_2)$ by its complex conjugate. The term $\mathbf{P}^{(2)}$ is also responsible for *second harmonic generation* accompanied by the appearance of double frequencies, $2\omega_1$ and $2\omega_2$, in the emission spectrum.[5] When three waves participate in the nonlinear interaction with the medium, then besides second-order effects, the processes of *third harmonic generation* and *four-wave mixing* can also be observed. They are described by the polarization $\mathbf{P}^{(3)}$ and lead to the generation of waves with frequencies equal to different combinations $\pm\omega_i \pm \omega_j \pm \omega_k$, where the indices i, j and k run over the values $1, 2, 3$.

Nonlinear optical effects have maximum efficiency if the *phase matching conditions* are satisfied. These conditions follow from the δ-functions relating the wave vectors (Eq. (6.8)). They express the conservation of momentum parallel to the surface plane and read as

$$\mathbf{k}_\parallel = \mathbf{k}_{1\parallel} + \mathbf{k}_{2\parallel} \qquad (6.9)$$

and

$$\mathbf{k}_\parallel = \mathbf{k}_{1\parallel} + \mathbf{k}_{2\parallel} + \mathbf{k}_{3\parallel} \qquad (6.10)$$

for sum-frequency generation and four-wave mixing, respectively.

Nonlinear optical susceptibilities

The nonlinear polarizations, $\mathbf{P}^{(n)}$, in the expansion (6.6) can be represented in a form similar to the sum (6.8), where each term describes a nonlinear optical process specified by the relevant δ-functions. It is reasonable, therefore, to write the polarization corresponding to a specific process as

$$\mathbf{P}^{(n)}(\omega = \omega_i + \omega_j + \ldots) = \chi^{(n)}(\omega; \omega_i, \omega_j, \ldots) : \mathbf{E}(\omega_i)\mathbf{E}(\omega_j)\ldots \qquad (6.11)$$

where we have taken into account the δ-function for frequency and have omitted the wave vectors for brevity. For example, sum-frequency generation is then described by the term

$$\mathbf{P}^{(2)}(\omega_1 + \omega_2) = \chi^{(n)}(\omega_1 + \omega_2; \omega_1, \omega_2) : \mathbf{E}(\omega_1)\mathbf{E}(\omega_2) \qquad (6.12)$$

whereas second harmonic generation is described by

$$\mathbf{P}^{(2)}(2\omega) = \chi^{(n)}(2\omega; \omega, \omega) : \mathbf{E}(\omega)\mathbf{E}(\omega) \qquad (6.13)$$

The nonlinear optical susceptibilities can be calculated, in principle, on the basis of the density matrix formalism. However, one can often draw some conclusions about the nonlinear optical output from symmetry considerations. The nonlinear susceptibility tensors reflect the structural symmetry of the crystal since they are determined by its electronic or vibrational states.

---

5) Note that second harmonic generation occurs also when an intense *single* wave strikes the surface or interface.

As a consequence, some tensorial elements are equal to zero and others are related to each other, thus greatly reducing the number of independent elements. For example, in general, the second-order susceptibility tensor $\chi_{ijk}^{(2)}$ has in total 27 elements which are reduced to four nonzero independent elements for a hexagonal crystal of symmetry $6mm$ and to a single one for a cubic crystal of symmetry $\bar{4}3m$.

Surface versus volume contributions

In a similar way to linear optics, the question about the relative contribution of a surface or an interface as compared with the overall signal intensity is crucial in nonlinear optics. In a simplified approach, the surface region is represented by a microscopic layer of thickness d with optical constants that differ from those in the bulk. Since linear optical phenomena are usually determined primarily by bulk properties, we can assume that the surface layer has the bulk refractive index. However, the nonlinear susceptibilities of the surface, χ_s^{nl}, and of the bulk, χ_b^{nl}, can differ significantly from each other. If the effective length of nonlinear optical interaction in the bulk is equal to l, the bulk contribution to the signal is estimated to be $|\chi_b^{nl}|l$, whereas that of the surface is determined by $|\chi_s^{nl}|d$. The ratio between these two quantities depends on different factors. The surface contribution can be essentially enhanced if at least one of the incident waves is in resonance with a transition between surface energy levels. On the other hand, the bulk contribution can be suppressed if the length l is limited due to a large phase mismatch expressed by the conditions (6.9) or (6.10). In addition, due to different symmetry properties, some tensorial components of the nonlinear susceptibility can be equal to zero in the bulk, while they are nonzero at the surface. In this case one can select the surface contribution by an appropriate choice of input and output polarizations. This is also possible for thin films deposited onto a surface.

A special case where the surface contribution becomes dominant is realized for nonlinear optical effects of even orders (second order, in particular). If a medium possesses inversion symmetry, its optical properties must not be changed on the inversion operation: $\mathbf{r} \rightarrow -\mathbf{r}$. On the other hand, both $\mathbf{E}$ and $\mathbf{P}$ are the polar vectors, i.e., they change their sign on inversion. From Eq. (6.1) we conclude that both properties are compatible only if all even-order terms are identically equal to zero. This is not the case in the surface region of the crystal where the inversion symmetry is broken. This finding allows one to use second harmonic and sum-frequency generation as surface/interface specific optical tools at surfaces of centrosymmetric crystals.

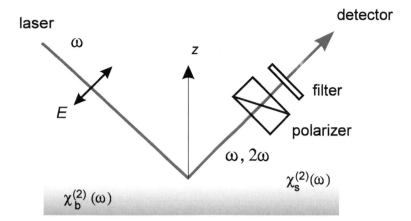

Fig. 6.1 Setup for the observation of optical frequency doubling at surfaces with surface susceptibility $\chi_s^{(2)}(\omega)$. The solid has the bulk susceptibility $\chi_b^{(2)}(\omega)$. The filter solely transmits the frequency-doubled light. The arrow drawn perpendicular to the line of the incoming laser beam characterizes the direction of polarization (p-polarized, i.e., parallel to the plane of incidence). Reprinted from Rubahn (1999). Copyright 1999, with permission from John Wiley & Sons Limited.

6.2
Second Harmonic Generation

Figure 6.1 shows the setup for an experiment to observe optical second harmonic generation, SHG, at surfaces.[6] A pulsed laser beam of wavelength 1064 nm (fundamental of a Nd:YAG laser) irradiates the substrate under an angle of 45° with respect to the surface normal and is reflected. The generated SH signal at 532 nm is separated from the fundamental light by use of a monochromator and a filter and is detected by a photomultiplier. The polarization direction of the incoming laser beam is either parallel (p-polarized) or perpendicular (s-polarized) with respect to the plane of incidence. The SH signal too might be observed as p- or s-polarized using a polarizer. The restriction to defined polarization combinations means that only selected components of the $\chi^{(2)}$ tensor contribute to the signal intensity and allows one to deduce information about symmetries at the surface (see below).

As detailed above, a necessary condition for the generation of nonlinear polarization of even order is the lack of inversion symmetry in the nonlinear medium. In media with inversion symmetry, only nonlinear terms of odd order exist (beginning with $\chi^{(3)}$) as well as higher order bulk contributions (magnetic dipole, electric quadrupole, etc.). These terms are usually small compared with the dipole term.

---

6) A detailed account of the principles of surface SHG is given in
   Brevet (1997).

At the surface, however, the inversion symmetry of the bulk is broken. Electric dipole contributions to the nonlinear polarization become possible due to the spatial structure of the surface and due to the discontinuity of the normal component of the electric field at the surface. In the case of a *nonlocal* interaction, the polarization at a given position $\vec{r}$ depends on the external field of the surroundings (i.e, "spatial dispersion"). In most cases, however, one assumes for the sake of simplicity a *local* interaction, in which case the susceptibility is independent of the polarization in the surroundings. If one assumes that the main reason for this nonlinear polarization is the generation of a strong (static) dipole field at the surface, then it becomes clear that it should be localized in the uppermost atomic layers down to a depth of 0.5–1.5 nm (Sipe et al. 1987).

In the case of metals, the damping of the incoming electromagnetic light wave by Friedel oscillations provides the most important length scale (Song et al. 1988). The wavelength of the oscillations is π/k_F with the Fermi wavevector (the largest possible wavenumber of the free electron gas) $k_F = 2\pi/\lambda_F = (3\pi^2\bar{n})^{1/3}$ and $\bar{n}$ the average density of the positive charge of the ion cores. If one uses for $\bar{n}$ the value for gold (5.9×10^{22} cm^{-3}), then one obtains $k_F = 1.2$ Å^{-1} and $\lambda_F = 5.24$ Å. The damping thus mainly occurs over a depth of a few nanometers. At the interface with the vacuum the electronic charge density is smeared out. This so-called "spill out" of the electrons leads to an electrostatic dipole layer.

The additional electric quadrupole and magnetic dipole contributions from the bulk can be separated from the real surface contributions only under special conditions: for example, if one modifies the surface layer in the form of an evaporated thin film and if one extracts the bulk contributions by calculating the interface contributions (Koopmans et al. 1993). The bulk contributions are proportional to the field gradient, meaning that a zone of about $\lambda/2\pi \approx 5\ldots10$ nm contributes to the total signal intensity.

6.2.1
Determination of Coverages

If one takes the above discussed restrictions into account, SHG is, for media with inversion symmetry, an extremely surface-sensitive method. This class of media covers face centered cubic, fcc (gold, etc.) and body centered cubic, bcc (sodium, etc.) metals, diamond, silicon, germanium, all gases and liquids or glasses. Depending on the absolute value of the components of $\chi_S^{(2)}$, adsorption and desorption processes might be observed in the submonolayer regime. This is demonstrated in Fig. 6.2 for CO molecules, adsorbed on Ni(111) (Zhu et al. 1985) and in Fig. 6.3 for alkali films on Si(111) (Bordo et al. 2005). The

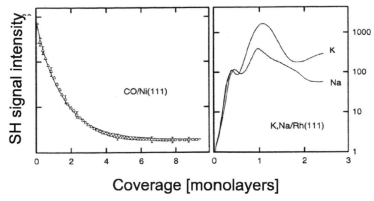

Fig. 6.2 Coverage dependence of the intensity of frequency-doubled light, induced by irradiating a Ni(111) surface with 1064 nm laser light. The surface has been covered with CO molecules. Reprinted from Zhu et al. (1985), Copyright 1985, with permission from Elsevier Science.

signal intensity is given by Shen (1986)

$$I_{2\omega} = \frac{32\pi^3\omega \sec^2\Theta_i}{c^3\hbar\epsilon(\omega)\sqrt{\epsilon(2\omega)}} \cdot |\chi_s^{(2)}|^2 I_\omega^2 \cdot F \cdot \tau \tag{6.14}$$

Here, Θ_i is the angle of the incoming laser beam with respect to the surface normal. For a Nd:YAG laser pulse (1064 nm) of duration $\tau = 10$ ps, intensity $I_\omega = 10^9$ W/cm^2 and area $F = 1$ mm^2 (corresponding to a pulse energy of 100 μJ), which irradiates a monolayer of molecules with a surface susceptibility of $\chi_s^{(2)} = 10^{-15}$ esu, one calculates from Eq. (6.14) about 10^5 photons per laser pulse. In the CGS system the susceptibility is defined via $P^{(n)} = \chi^{(n)}E^{(n)}$. The conversion factor between the two systems is $\chi^{(n)}[(\text{cm/V})^{n-1}] = \chi^{(n)}[\text{esu}] \cdot 4\pi/(3\times10^2)^{(n-1)}$ (Butcher and Cotter 1990). For the presented example this means $\chi^{(2)} \approx 4\times10^{-17}$ cm/V. This value is six orders of magnitude smaller compared with that for a typical nonlinear crystal such as BBO. This signal intensity can be measured without difficulty using a photomultiplier. The total susceptibility of the monolayer is given by the nonlinear polarizability of each single molecule, $\alpha^{(2)}$, and the number density of the surface molecules, N_s,

$$\chi_s^{(2)} = N_s \cdot \alpha^{(2)} \tag{6.15}$$

if one averages over the orientations of the molecules and neglects their mutual interactions.[7]

---

7) Mutual interactions of the nonlinear optically active particles modify the local fields (Ye and Shen 1983).

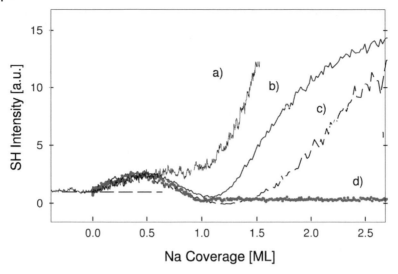

Fig. 6.3 Coverage dependence for an alkali-film-covered Si(111)(7×7) surface for different surface temperatures and excitation wavelengths. a) $T_S = 240$ K, $\lambda = 1067$ nm, b) $T_S = 150$ K, $\lambda = 570$ nm, c) $T_S = 150$ K, $\lambda = 497$ nm, d) $T_S = 326$ K, $\lambda = 570$ nm. The intensity from the plain Si surface is denoted by a dashed horizontal line. Reprinted with permission from Bordo et al. (2005). Copyright 2005, EDB Sciences.

For the surface coverage of a monolayer ($\approx 5 \times 10^{14}$ cm^{-2}) the value $\chi_S^{(2)} = 10^{-15}$ esu corresponds to a molecular polarizability of 2×10^{-30} esu. The nonlinear polarizabilities of metals are often more than an order of magnitude larger; those of dye-doped monofilms can be even higher. In Table 6.1 exemplary measured values of $\chi_s^{(2)}$ are summarized for different excitation wavelengths and the coverage of a monolayer. It is to be noted that bottom-up technologies such as supramolecular chemistry can result in molecular films that show a significantly higher nonlinear optical efficiency. For example, second order nonlinear susceptibilites of 10^{-11} esu have been demonstrated by generating Langmuir–Blodgett films of nested chiral molecules (Verbiest et al. 1998).

Upon reflection from the surface, the local field strength tensor **L** has to be taken into account and the above $\chi_s^{(2)}$ has to be modified accordingly:

$$\chi_s^{(2)} \rightarrow \mathbf{L}(2\omega)\chi_s^{(2)}\mathbf{L}(\omega)\mathbf{L}(\omega) \tag{6.16}$$

In the case of an ideal smooth surface, **L** is mainly represented by Fresnel factors (Mizrahi and Sipe 1988). These factors depend strongly on the angle of incidence of the light and average over the optical properties via a general dielectric function of the medium down to the penetration depth of the light. Hence they are not surface sensitive in the same way as is $\chi^{(2)}$. For a rough surface, the electromagnetic field enhancement might be much stronger

Tab. 6.1 Selected nonlinear surface susceptibilities. The values are from: a, Wang and Duminski (1968); b, Heinz et al. (1982); c, Kelly et al. (1991); d, Chen et al. (1981); e, Bloembergen et al. (1968); f, Chen et al. (1973); g, Marowsky et al. (1988).

Material	$\chi_s^{(2)}$ [10^{-17} esu]	λ [nm]	Reference	Remark
Lithiumfluoride	5	694	a	insulator
Rhodamine 110	150	670	b	dye
Silicon	160	1064	c	semiconductor
Pyridine	800	1064	d	dye
Silver	900	1064	e	metal
Sodium	2700	694	f	metal
Hemicyanine film	50000	1064	g	fatty acid film

due to resonance effects. For example, for pyridin molecules adsorbed onto a rough silver electrode, the effective surface susceptibility is about a factor of 50 higher compared with that for a smooth surface, namely $\chi_S^{(2)} \approx 4 \times 10^{-13}$ esu (Boyd et al. 1986). This strong nonlinearity makes the measurement of SH signals, even with continuous lasers and a power of a few tens of milliwatts, possible; the good spatial resolution of the laser beam (focal diameter a few μm) allows one to perform nonlinear surface microscopy via SH generation (Boyd et al. 1986; Smilowitz et al. 1997). For example, by exploiting magnetic field influences on SHG (see later in this chapter) magnetic domains could be imaged via SHG microscopy (Kirilyuk et al. 1997b, c).

Figures 6.2 and 6.3 show changes in the SH signal intensity from metal and semiconductor surfaces as the surfaces are covered by CO molecules and sodium atoms, respectively. The SH signal intensity was induced by 60 fs pulses at 570 nm (solid line, circles) and at 497 nm (dashed line) in 60° reflection geometry and in pp-orientation.

Obviously the SH intensity of the plain Ni(111) surface decreases with increasing coverage by CO molecules since CO is an electron acceptor, which binds the initially nearly free electrons of the nickel surface. In contrast, alkali atoms are electron donors and thus increase the magnitude of the nonlinear surface signal intensity (Tom et al. 1986). Hence the initial increase in SH intensity with coverage for $\Theta < 0.5$ ML is dictated by the generation of Na/Si induced surface dipoles and thus is independent of the surface temperature and also of the wavelength of the fundamental beam. A further increase in coverage leads to a depolarization of the dipoles due to mutual interactions and an accompanying decrease in the binding energy between Na and Si, which is accompanied by a structural transformation of the Na overlayers, since above the $\Theta = 0.5$ ML the lattice sites for covalent bindings are occupied.

At room temperature, the strong decrease in the sticking probability for Na on Si(111) (Papageorgopoulos and Kamaratos 1992) prevents a subsequent increase in signal intensity. At significantly lower surface temperatures the following increase in SH signal intensity is due to collective electronic excitation since the adsorbate layer becomes metallic at a coverage close to 1 ML (Soukissian et al. 1989), corresponding to the formation of a conduction band within the Na overlayer (Jeon et al. 1992). With increasing alkali coverage the surface electron density increases strongly, shifting the surface plasmon frequency of the adsorbate from the value of the 3S–3P transition for free Na atoms ($\hbar\omega_{3S-3P} = 2.1$ eV) towards the value of the bulk-terminated sodium surface ($\hbar\omega_{sp} \approx 4.1$ eV) at coverages of the order of 2 ML (Jostell 1979). Hence a near-resonance enhancement of the nonlinear signal intensity is expected since $\hbar\omega_{sp}$ approaches the value of the second-harmonic frequency, $2\hbar\omega = 4.35$ eV. If one increases the fundamental beam frequency ω and consequently 2ω, then the slope of the signal increase is reduced due to the enlarged difference between 2ω and ω_{sp}.

Figure 6.2 also demonstrates the possibility of SHG to discriminate between different growth modes of ultrathin films on surfaces, namely layer growth (at room temperature) and "Stranski–Krastanoff" growth (layer growth, followed by three-dimensional island growth). If one exchanges the substrate and uses, for example, a dielectric instead of the semiconductor surface, then alkali films will show a three-dimensional island growth mode ("Volmer–Weber" growth mode) from the beginning. The measured SH intensity from an array of spherical islands is given by Chen et al. (1983)

$$I(2\omega) \propto I^2(\omega)|\chi^{(2)}|^2 Q^2(\omega)Q(2\omega)r_0^4 \tag{6.17}$$

and thus depends on the cluster radius r_0. The local field tensors **L** in Eq. (6.16) are represented by the factors $Q(\omega)$, which describe the mean local electric field at the cluster surface in units of the geometrical cross-section:

$$Q(\omega) = \frac{R^2}{\pi r_0^2} \int_0^{2\pi} \int_0^{\pi} E_{\text{int}} E_{\text{int}}^* \sin\Theta \mathrm{d}\Theta \mathrm{d}\Phi \big|_{R=r_0} \tag{6.18}$$

The validity of this approach can be verified by growing alkali islands on dielectrics such as mica or lithium fluoride. As demonstrated in Fig. 6.4 by comparison of calculated and measured (Müller et al. 1997) relative SH intensities as a function of cluster radius, the coverage-dependent SH signal intensity shows pronounced resonances at well-defined values of mean cluster radius that depend on the excitation wavelength. These resonances are the nonlinear optical analog to the resonances observed in linear extinction spectra of cluster films and can be predicted via classic Mie theory by resonances in the local field factors. They in turn are due to a collective electronic, surface plasmon excitation. Thus SHG in this case provides both a unique fingerprint of the Volmer–Weber growth mode and a means of determining the mean cluster radius.

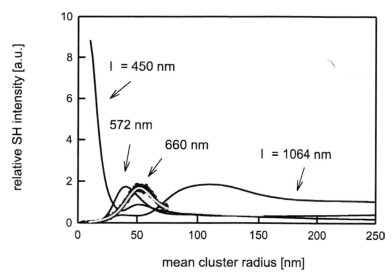

Fig. 6.4 Calculated relative second harmonic intensities as a function of fundamental wavelength λ and mean cluster radius for sodium clusters adsorbed on dielectrics. Typical experimental values are indicated by dots. Reprinted from Rubahn (1999). Copyright 1999, with permission from John Wiley & Sons Limited.

In the past, strongly enhanced second-harmonic signals due to local field enhancement have been observed for thin alkali films in the ATR (attenuated total reflection) geometry (Simon et al. 1975) and later also in reflection geometry on rough silver and gold films (Wokaun et al. 1981; Chen et al. 1981). Finding a strong relationship between enhanced SHG and enhanced Raman scattering (SERS) (Boyd et al. 1984), a phenomenological treatment of the problem for small spherical particles (Chen et al. 1983) was provided. More accurate treatments using Green's function formalisms (Hua and Gersten 1986) and nonlinear Mie scattering (Östling et al. 1993; Dewitz et al. 1996) were applied subsequently, permitting a qualitative prediction of the SH enhancement even for larger spheres. So far, in all theoretical investigations of resonance enhancement, an accurate calculation of $\chi^{(2)}$ from an electronic theory has not been provided. As a consequence, $\chi^{(2)}$ is always assumed to be independent of cluster size and the clusters are assumed to be noninteracting spheres (with the exception of calculations in (Garcia-Vidal and Pendry 1996)). These assumptions have also been made for the calculations that led to Fig. 6.4.

8) For flat metallic surfaces nonlocal response theory has shed some light on the resonance mechanism (Liebsch 1989; Jensen et al. 1997). However, it has been shown that due to a large contribution of bulk SHG to the total SH signal from adsorbed clusters, the SHG from metal films and cluster films cannot be easily compared (Aussenegg et al. 1995).

The sensitivity of the method increases for given laser energy if one uses higher irradiances or shorter pulses, since it follows from Eq. (6.14) due to the proportionality to I_ω that

$$I_{2\omega} \propto I_\omega^2 A \tau \propto \frac{1}{\tau} \qquad (6.19)$$

with τ being the pulse duration and A the irradiated area. Limits are given by melting of the surface, which usually occurs for pulsed excitation with fluences of the order of J/cm^2. For excitation with continuous lasers the thermal diffusion coefficient of the surface (the time constant at which heat is transported out of the radiation zone) determines the maximum power that can be applied (about 1 W for a strongly focused laser beam on silicon).

The SH method for determining surface coverages has the same sensitivity as methods using linear reflectivity changes and it has no major restrictions regarding the observable adsorbates. However, while coverage changes can be deduced straightforwardly from the measured SH values, the evaluation of *absolute* coverages is only possible by the use of reference measurements (e.g., thermal desorption measurements) (Zhu et al. 1985) or Auger measurements.

6.2.2
Symmetries

The $\chi^{(2)}$ tensor reflects the symmetries of the surface via the symmetry properties of the electronic surface states. Since it is a tensor of rank three, only three- or lower-fold surface symmetries can be resolved in dipole approximation ($L = 1$). If it is possible to measure higher order multipole-contributions, then the highest resolvable rotational symmetry is for an Nth-order nonlinear technique given by $(N + L)$ (Koopmans et al. 1992).

According to Eq. (6.13) the frequency dependence of the second-order nonlinear component of the induced polarization is given by $P_l(2\omega) = \sum_{m,n} \chi_{lmn}^{(2)}(2\omega; \omega, \omega) E_m(\omega) E_n(\omega)$, where $\chi_{lmn}^{(2)} = |\chi_{lmn}^{(2)}| e^{i\phi_{lmn}}$ couples the mth and nth components of the fundamental with the lth components of the generated nonlinear polarization (ϕ_{lmn} is the phase). The $\chi^{(2)}$ tensor, due to the condition $\omega_1 = \omega_2 = \omega$, might be contracted to 18 components ("piezoelectric contraction"), which – depending on the symmetry of the surface – are not independent of each other (Yariv 1989). If the exciting laser irradiates along the surface normal (z-direction), then the emitted SH intensity is the following function of the angle Θ between the electric field vector of the linearly polarized laser and the preferred directions on the crystal (labelled x and y) (McGilp 1987),

$$I_x(2\omega) \propto |(\chi_S^{(2)})_{xxx} \cos^2 \Theta + (\chi_S^{(2)})_{xyy} \sin^2 \Theta + (\chi_S^{(2)})_{xyx} \sin 2\Theta|^2 \qquad (6.20)$$

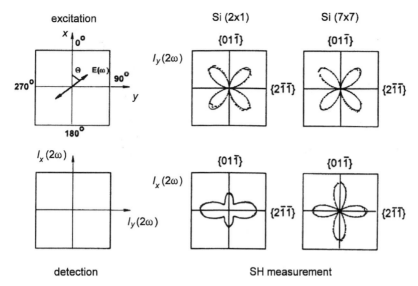

excitation   Si (2x1)   Si (7x7)

detection   SH measurement

Fig. 6.5 Symmetries of the reconstructed silicon (2×1) and (7×7) surfaces, observed via rotation of the polarization vector of the exciting laser (field strength $E(\omega)$) within the crystal plane and measurement of the second harmonic intensity $I(2\omega)$ along the $(2\overline{1}\overline{1})$ and $(01\overline{1})$ crystallographic directions, respectively. Reprinted with permission from Heinz et al. (1985). Copyright 1985, American Physical Society.

and

$$I_y(2\omega) \propto |(\chi_S^{(2)})_{yxx}\cos^2\Theta + (\chi_S^{(2)})_{yyy}\sin^2\Theta + (\chi_S^{(2)})_{yxy}\sin 2\Theta|^2 \qquad (6.21)$$

For laser irradiation along the surface normal one expects SH signals by the symmetry classes 1, $1m$ and 3, $3m$.[8] For $1m$-symmetry the nonlinear susceptibilities vanish in the directions xyx, yyy and yxx, and thus in the y-direction the observed intensity I_y is proportional to $\sin^2 2\Theta$. In Fig. 6.5 the measured SH intensities along the y-direction (upper part) and x-direction (bottom part) are shown for two differently reconstructed silicon crystals: the (2×1) reconstruction, which results from cleaving of the single crystal in vacuum, and the (7×7) equilibrium reconstruction, which can be obtained by annealing to surface temperatures of, usually, above 1000 °C. The solid lines in the y-direction are fits to the measured points assuming a $1m$-symmetry.

In the x-direction one expects to observe in the case of the higher $3m$-symmetry $(-xxx = xyy = yxy; xyx = 0)$ an intensity dependence proportional to $\cos^2 2\Theta$, since $\cos^2\Theta - \sin^2\Theta = \cos 2\Theta$. Experimentally, this intensity dependence is observed only for the (7×7) reconstructed surface,

8) This corresponds in the "Schönflies notation" (Schönflies 1891) to the crystal classes C_1 (single rotation axis, tricline), C_s (single mirror plane, monocline), C_3 (triple rotation axis, rhombohedric) and C_{3v} (triple rotation axis and three vertical mirror planes, rhombohedric) (Hamermesh 1962).

while the (2×1) surface shows a dependence in the x-direction, which can be reproduced via two independent tensor elements xxx and xyy. Hence the SH measurements directly prove that the (2×1) surface shows the lower $1m$-symmetry, while the (7×7) surface has three mirror planes.

The deduced symmetries are consistent with electron diffraction (LEED) results. In contrast to LEED studies, however, which do not allow one to exclude the existence of a higher order (e.g., sixfold) symmetry, this is possible via SH measurements since at normal incidence a surface with such high symmetry would show no SH signal intensity at all.

Via SHG it is possible to see the heat-induced increase in symmetry of the silicon surface if one observes the vanishing of the signal in a direction in which only the (2×1) surface induces SH signal intensity (Fig. 6.5). It can be seen that the phase transition from the lower to the higher symmetry has already started around 550 K surface temperature (Heinz et al. 1985).

Measurement of the rotationally anisotropic second harmonic yield has been shown to be a versatile tool for a nonintrusive, relatively simple determination of interface symmetry. Besides plain surfaces it, of course, has also been applied to the determination of adsorbate symmetries, which turns out to be a difficult task in the case of adsorption on insulating surfaces. For example, the adsorption symmetry of molecular water on alkaline-earth halides such as CaF_2 has been deduced as a function of coverage (Zink et al. 1992), demonstrating oriented initial adsorption. This orientation is subsequently lost for higher coverages.

In the case of an isotropic distribution in the (x, y)-plane, no rotational anisotropies are observed in the SH plots at normal incidence. However, if one irradiates with the laser under a fixed angle of incidence Θ_{in} with respect to the surface normal, for a given polarization of the electric field vector with respect to the plane of incidence, and varies the angle α of a given crystallographic direction on the surface (i.e., by azimuthal rotation of the sample with respect to the surface normal) (Jordan et al. 1995), then one is able to deduce different components of the effective $\chi^{(2)}$ tensor. The information content increases further if one also varies the angle of incidence (Bratz and Marowsky 1990; Ying et al. 1993).

6.2.3
Orientations and Chirality

The orientation of molecular adsorbates on surfaces might also be deduced from the polarization dependence of the SH signal (Andrews and Hands 1996). Let us assume that a monomolecular film has been adsorbed on the surface. The transition dipole moments of the molecules, which are responsible for the SH generation, define a preferred axis in the molecules, which is tilted with respect to the surface normal by the angle Θ. Let the molecules be isotropically distributed in the (x, y)-plane. Their microscopic polarizability in the direction of the molecular axis, z', $\alpha^{(2)}_{z'z'z'}$, determines the nonlinear optical response. Then the orientation angle Θ can be determined from a measurement of the two macroscopic tensor components (zxx) and (zzz) (Heinz 1991):

$$\sec^2 \Theta \approx 2\frac{\chi^{(2)}_{zxx}}{\chi^{(2)}_{zzz}} + 1 \qquad (6.22)$$

For this to hold, a knowledge of the absolute values of $\chi^{(2)}$ or of the surface density of the molecules is not necessary, assuming that the influence of local fields on the surface (especially the interaction between the molecules) can be neglected.

Especially for the application to biological relevant surfaces (e.g., adsorbed proteins) the sensitivity of polarized SHG to the chirality of the molecules (the "handedness" of their structures) is important (Verbiest et al. 1998). Recently, the second-order nonlinear optical analog to circular dichroism and optical rotatory dispersion spectroscopy has been successfully developed (Yee et al. 1994; Byers et al. 1994).

6.2.4
Surface Magnetization

The investigation of the magnetic properties of surfaces, thin films and layered structures is of practical importance for the optimization of magnetic data storage techniques. With respect to fundamental research, the investigation of the exchange interactions between thin films of quantum well states, the spin flip dynamics or the growth of magnetic domain structures are especially interesting.

The magnetic properties of surfaces can be investigated by spin-polarized electrons (Feder 1985) or via the magnetooptic Kerr effect (SMOKE, surface magnetooptic Kerr effect) (Qian and Wang 1990). These linear techniques suffer from their large penetration depth (about 50 nm in the case of SMOKE), meaning that their surface sensitivity is provided solely by the properties of the investigated material, e.g., the thickness of the evaporated film. The magnetism of interfaces between layers made of different materials ("buried interfaces") cannot be investigated independently of the magnetic properties of the bulk of the layers.

Since magnetization does *not* break the inversion symmetry of the investigated material ($\vec{M}$ is an axial vector with even parity with respect to the inversion operation), one does expect to preserve the large surface sensitivity of SHG even in the case of magnetic interfaces. On the other hand, the existence of the magnetization changes the symmetry of the bulk and surface. Thus ad-

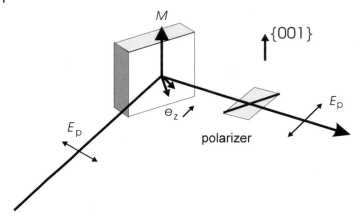

Fig. 6.6 Setup for the investigation of the influence of surface magnetization on the generation of frequency-doubled light from an Fe(110) crystal (light in pp-orientation). The symbol e_z denotes the Lorentz-force-induced rotation of the electron oscillation perpendicular to the surface. Reprinted with permission from Reif et al. (1991). Copyright 1991, American Physical Society.

ditional elements of the electric dipole tensor $\chi_s^{(2)}$ become nonzero compared with the nonmagnetized surface (Pan et al. 1989). Of special interest are the odd elements that change sign as a result of changing the direction of $\vec{M}$. This results in a change in the total SH-intensity

$$I_{+,-} \propto |\chi^{(2)}_{\text{nonmagnetic}} \pm \chi^{(2)}_{\text{magnetic}}|^2 \tag{6.23}$$

Depending on the direction of $\vec{M}$, the intensity should increase or decrease. Hence magnetic-field-induced SHG (MSHG) exists only if the surface is SH active even without a magnetic field, and $\chi^{(2)}_{\text{magnetic}}$ depends linearly on the magnetization. However, the nonlinear Kerr rotation might be much stronger compared with the linear one, as demonstrated recently by the use of a multilayer of 2 nm Fe on 2 nm Cr on a quartz substrate, which possesses only a very small linear Kerr rotation (Koopmans et al. 1995).

In a classic macroscopic picture, the nonlinear Kerr effect can be explained as follows. The reflection of an incoming light wave at the interface is related to the nonlinear generation of an electron flux which is affected by the surface magnetization via spin–orbit coupling and exchange interaction. The resulting Lorentz force $\vec{F} = -e\vec{v} \times \vec{B}$ acts perpendicularly to the magnetic induction and the velocity vector of the electrons and thus rotates the direction of the emitted electromagnetic field vector with respect to the incoming light vector. This is shown in Fig. 6.6 for p-polarized light, which hits an Fe(110) crystal that possesses a magnetization in the {001}-direction.

The projection of the incoming electric field vector induces an electron oscillation perpendicular to the surface, which is rotated counterclockwise by

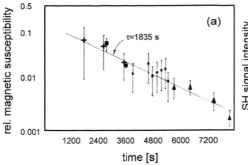
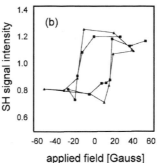

Fig. 6.7 (a) Relative magnetic surface susceptibility $\chi^{(2)}_{mag.}/\chi^{(2)}_{nonmag.}$ as a function of CO coverage. Reprinted with permission from Reif et al. (1991). Copyright 1991 American Physical Society. (b) Magnetic-field-induced SHG from a 50 nm cobalt/5 nm gold on a quartz substrate. Reprinted from Spierings et al. (1993), Copyright 1993, with permission from Elsevier Science.

the magnetic field. If the magnetization is directed in the negative {001}-direction, then the generated SH light will be rotated clockwise. If one now observes p-polarized frequency-doubled light along a fixed direction, then this magnetic-field-induced rotation results in a decrease or an increase in the signal intensity, depending on the direction of the magnetic field. This behavior has indeed been observed in the course of pp-measurements under 45° angle of incidence ($\lambda = 532$ nm, $\tau = 6$ ns) from a clean iron crystal, which has been mounted in an ultrahigh vacuum between the pole faces of an electromagnet (Reif et al. 1991). The observed ratio between magnetic-field-induced SHG and nonmagnetic SHG was about 0.25. This large effect makes the method a promising tool for other systems too. The surface sensitivity is demonstrated by observing that the signal intensity decreases exponentially with increasing coverage with background gas (mainly dissociative adsorption of CO) (Fig. 6.7a).

In Fig. 6.7b the sensitivity of the method to buried interfaces is further demonstrated by a measurement on a cobalt/gold/quartz layered system (Spierings et al. 1993). The magnetic-field-induced effect is observed *only* if a single (or an odd number of) cobalt/gold interfaces does exist and in that case shows a significant hysteresis. For a longitudinally magnetized PtMnSb(111) surface, circular dichroism[9] has been observed via SHG, which provides information about the population of spin-up and spin-down states in the conduction band (Reif et al. 1993).

9) In order to perform this experiment one irradiates the surface with right- or left-hand circularly polarized light, corresponding to photon spins oriented parallel or antiparallel with respect to the magnetization direction. The different SH yields for the different directions of rotation, result in a change in ellipticity of the nonlinearly reflected light as compared with the incoming light.

The versatility of the MSHG method is mainly due to its high sensitivity, which results partially from local-field enhancement effects.[10] Consequently, the method has found a large number of applications, including the investigation of quantum well oscillations (Wierenga et al. 1995; Kirilyuk et al. 1997a) or femtosecond time-resolved spin dynamics (Scholl et al. 1997; Hohlfeld et al. 1997).

6.2.5
Spectroscopy

The spectroscopic possibilities of SHG lie in the resonance enhancement of the signal, namely if the second harmonic frequency, 2ω, coincides with the frequency of a dipole-allowed transition ω_{fi} in the irradiated material. Here, "i" means the ground state and "f" an excited state. Possible resonance enhancement can be directly deduced from the microscopic expression[11] for $\alpha^{(2)}$:

$$\alpha^{(2)}(2\omega) \propto \sum_{k,f} \frac{M_{if}M_{fk}M_{ki}}{(2\omega - \omega_{fi} - i\gamma_{fi})(\omega - \omega_{ki} - i\gamma_{ki})} \tag{6.24}$$

where M denotes the matrix element for dipole transitions between two states ("k" means a real intermediate state) and γ denotes the half-width of the transition, corresponding to the characteristic relaxation time (Shen 1984). Obviously the hyperpolarizability is increasing strongly if the difference between 2ω and ω_{fi} or ω_{ki} becomes small ("resonance"). Note that at the spectral position of resonance the Taylor expansion of the induced polarization might no longer be valid and the transition probability is dominated by additional dynamic effects such as photon echos, free induction decay, self-induced transparency, etc. (Mandel and Wolf 1995). It is thus difficult to model quantitatively the absolute signal intensity at resonance.

Figure 6.8 shows the measured SH signal from a silicon (100) surface, which has been covered by a 700 nm thick SiO_2 layer, as a function of laser energy. A prominent resonance is seen at 3.3 eV. Adsorption of oxygen on the surface reduces the height of the resonance insignificantly. This means that the signal is not generated by the surface states of silicon. However, comparison of the energy dependence of the linear susceptibility $\chi(2\omega) = \epsilon(2\omega) - 1$ of bulk silicon (solid line) with the measured data, shows good agreement concerning the positions of the resonance. Hence the resonance might be induced by the direct valence–conduction band transition in silicon. Since the signal can be

---

10) Obviously, the sensitivity can be further enhanced by inducing the second-harmonic signal in the spectral neighborhood of a surface plasmon resonance of the thin magnetic film.

11) This equation describes the molecular hyperpolarizability via second-order perturbation theory assuming a single-particle excitation. Collective phenomena are not taken into account.

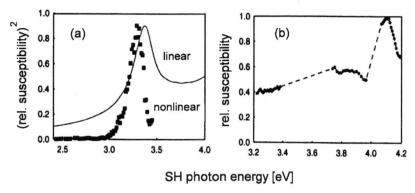

Fig. 6.8 (a) Measured nonlinear susceptibility of second order of Si(100)/SiO₂. The solid line is the calculated linear susceptibility. Reprinted with permission from Daum et al. (1993). Copyright 1993, American Physical Society. (b) Measured nonlinear susceptibility for Cu(111) using angle of incidence $\Theta = 67°$ and azimuthal angle $\psi = 30°$. Reprinted with permission from Lüpke et al. (1994). Copyright 1994, American Physical Society.

observed even from below a 700 nm thick silicon oxide layer, it is probably generated at the interface between the silicon and SiO₂, which does not possess inversion symmetry. The slight red shift of the measured as compared to the calculated resonance provides evidence that the Si–Si bindings are slightly elongated at the surface as compared with the bulk.

In contrast to the 3.3 eV resonance at the silicon surface the 4.1 eV resonance on a Cu(111) surface (Fig. 6.8b) reacts sensitively on the adsorption of oxygen. This is a hint that, in that case, transitions into *surface states* are responsible for the enhancement of the nonlinear signal intensity. The most probable candidate is a transition between a surface state close to the Fermi energy and the ($n = 1$) image potential state. The energetic position of the resonance is $E = E_a - E_b = 4.92 - 0.82\,\text{eV} = 4.1$ eV. Here, E_a is the work function of Cu(111) and E_b is the binding energy of the image potential state. The long lifetime of the Rydberg-like image potential states increases the possible population density and thus the probability for the existence of nonlinear processes between them and the surface states.

Frequency-dependent SH measurements at Ag(110) have detailed different resonances in the energetic range between 1.6 eV and 2.1 eV (fundamental of the laser), which can be traced back partially to interband transitions and partially to transitions between occupied and nonoccupied surface states (Urbach et al. 1992). In the regime of interband transitions (for silver $2\omega = 3.8$ eV) the corresponding field enhancement results in an enhancement of the SH signal intensity. A similar effect occurs if 2ω corresponds to the transition between an occupied surface state below the Fermi level and an unoccupied state in the gap. Since transitions from bulk states of the same symmetry as the occupied surface state (but situated deeper below the Fermi level) might also occur, the resonance maximum usually has a broad tail to higher energies.

Up to now the experimentally observed resonances in the SH signals, which provide characteristic information about the electronic band structures and sometimes are truly surface-specific (i.e., they have no analog in the bulk, e.g., (Erley and Daum 1998)), could not be reproduced fully by theoretical approaches. Since the second-order nonlinear response of the surface to an incoming light wave reacts sensitively to the charge density profile of the ground state (Murphy et al. 1989), a detailed understanding of the electronic surface structure is important. SH theory, on the other hand, mostly uses a jellium model for the surface structure. It is not surprising that totally neglecting the lattice potential only in the case of surfaces with low electron densities and in the absence of surface defects (i.e., in the case of the homogeneous distribution of ions) provides qualitatively correct predictions of the nonlinear optical response function of the surface.

6.3
Sum Frequency Generation

An obvious disadvantage of using second harmonic generation for the spectroscopy at adsorbate-covered surfaces is the missing molecule specificity. This problem can be overcome if one mixes a fixed-frequency laser pulse with a pulse of variable frequency ("sum frequency generation", SFG). Since $\chi^{(2)}_{SFG}$ also vanishes in dipole approximation in media with inversion symmetry, this method has high surface sensitivity, too, and shows resonance enhancement.

Usually the second harmonic of a Nd:YAG laser ($\lambda_{vis} = 532$ nm) irradiates the surface at a given angle Θ_{vis} with respect to the surface normal (Fig. 6.9). This light is mixed with infrared light (ω_{IR}), which might have been generated in an optical parametric oscillator with variable wavelength. The sum frequency $\omega_{SF} = \omega_{vis} + \omega_{IR}$, which is generated in the substrate, is reflected under an angle Θ_{SFG} with respect to the surface normal, which is given by Hunt et al. (1987)

$$\omega_{SFG} \sin \Theta_{SFG} = \omega_{vis} \sin \Theta_{vis} + \omega_{IR} \sin \Theta_{IR} \tag{6.25}$$

The sum-frequency light might be separated from the irradiating light either by virtue of its different frequency or spatially. If one tunes the infrared light, then in the case of a resonance (e.g., by excitation of an adsorbate vibration at the surface) the nonlinear signal intensity is strongly increasing, just as in case of SHG (Eq. (6.24)). The enhanced molecule specificity of the method results from the use of IR light, which allows one to directly excite intramolecular motions. As an example, in Fig. 6.9, SFG spectra for methanol (CH_3OH) and

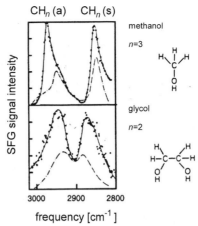

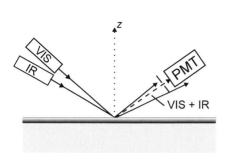

Fig. 6.9 Sum frequency generation at surfaces. On the left-hand side the experimental setup is shown schematically. An infrared (IR) and a visible (VIS) laser beam are mixed in a thin methanol adsorbate film and the resulting sum frequency signal is spatially separated and detected by a photomultiplier (PMT). If one tunes the wavelength of the IR laser, then adsorbate-spectroscopy can be performed (right-hand side). Reprinted from Hunt et al. (1987), Copyright 1987, with permission from Elsevier Science.

glycol ($C_2H_4(OH)_2$) are shown, which have been adsorbed on a quartz substrate (Hunt et al. 1987). The symmetric (s) and antisymmetric (a) CH stretch vibrations are clearly identified. For comparison, Raman spectra from the solution are shown by dashed lines. Their maxima are strongly shifted due to the mutual interactions of the molecules.

An experimental setup for the spectroscopy of silicon–hydrogen stretch vibrations via SFG is shown in Fig. 6.10. A mode-coupled Nd:YAG laser pumps a dye laser, which generates tunable 70 ps light pulses. Part of the YAG fundamental light (1064 nm) is amplified by a regenerative amplifier and is frequency doubled. This light then serves as the visible mixing pulse for the SFG, having 100 ps duration and 50 mJ energy. In addition, it is used also to amplify the tunable dye laser pulse to 6 mJ. The dye laser pulse is Raman-shifted to the infrared spectral regime (4.4–5.6 µm) inside a cell which is filled with cesium vapor. This procedure results in about 40 µJ tunable IR light. Both pulses together allow one to determine the energy of the Si-H stretch vibration at a smooth surface (2084 cm^{-1}). The setup shown in Fig. 6.10 can be used also to generate an IR pump pulse. In that way an adsorbate mode can be excited and the dynamics of the following relaxation, induced by the interaction with the substrate or with the surrounding adsorbed atoms, can be investigated by applying a temporal delay between pump and probe pulses. The identification of short-lived chemical species produced on a surface (Domen et al. 1998) but also at buried interfaces (Cremer et al. 1995) became possible via SFG vibrational spectroscopy.

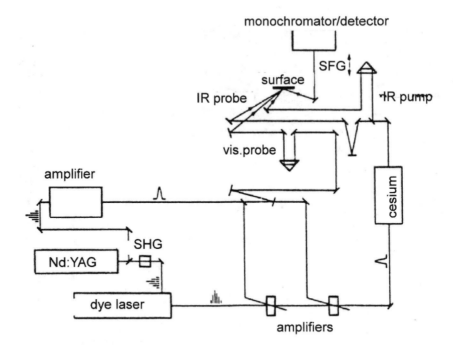

Fig. 6.10 Setup for IR pump/SFG probe, time-resolved nonlinear vibrational spectroscopy. Reprinted with permission from Morin et al. (1992). Copyright 1992, American Institute of Physics.

While the spectral position of the resonance can be determined straightforwardly using SFG, evaluation of the line profile with respect to the origin of possible broadenings or the absolute densities of the adsorbates becomes difficult. The basic problems with all nonlinear spectroscopic techniques (SHG, SFG, CARS, etc.) are: (i) a wavelength dependence of the local field strength tensors $\mathbf{L}(\omega, 2\omega)$ (Eq. (6.16)); and (ii) the nonresonant background, which is given by the additive nonresonant nonlinear susceptibility. Both factors have to be taken into account for an accurate profile analysis via extended fit procedures.

6.4
Four-wave Mixing

Optical second harmonic and sum frequency generation are three-wave mixing processes, which are intrinsically surface sensitive. The nonlinear polarization, which is necessary for the generation of the third photon, can be described by the second term in Eq. 6.1. If one takes additional photons into account, multi-wave mixing processes and especially four-wave mixing takes

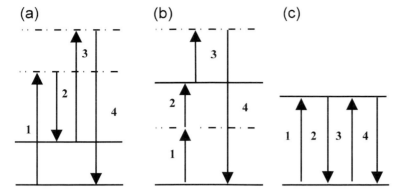

Fig. 6.11 Possible four-wave mixing processes. Real states are plotted by solid lines, virtual states by dash-dotted lines. 1,2,3 and 4 denote the frequencies ω_1 to ω_4. (a) Coherent anti-Stokes Raman scattering (CARS), $\omega_1 = \omega_3$. (b) Third harmonic generation (THG), if $\omega_1 = \omega_2 = \omega_3$, otherwise "up-conversion". (c) Degenerate four-wave mixing (DFWM), $\omega_1 = \omega_2 = \omega_3 = \omega_4$.

place. The coupling constant between irradiating fields and induced polarization then is the nonlinear susceptibility of third order, $\chi^{(3)}$. These methods are not intrinsically surface sensitive from symmetry arguments. However, if the signals are generated in thin adsorbate films or at (buried) interfaces they become interesting methods for surface investigations.

Figure 6.11 shows term schemes for some important four-wave mixing processes, which have been deduced by invoking energy conservation: coherent anti-Stokes Raman scattering (CARS), frequency tripling (THG) and degenerate four-wave mixing (DFWM).

As seen, the term scheme for DFWM is especially simple: all four photons induce transitions between real states. This "resonance enhancement" results in a higher signal strength as compared with, for example, CARS. Momentum conservation determines the direction of the resulting signal wave. Since the phase-matching condition is (for the notation see Fig. 6.13):

$$\mathbf{k}_{fp} + \mathbf{k}_{bp} - \mathbf{k}_{p} - \mathbf{k}_{s} = 0 \qquad (6.26)$$

$\mathbf{k}_{fp} + \mathbf{k}_{bp} = 0$ and the energies of all involved photons are equal, the phase conjugate signal wave will counterpropagate to the probe wave (phase-conjugate (PC) geometry; Fig. 6.12).

DFWM (Fisher 1983) is in fact a real-time variant of optical holography, which has been known since the sixties of the 19th century: two laser beams (here, forward pump and object beam) are overlapped coherently under a small angle Θ. They induce in a nonlinear medium with susceptibility $\chi^{(3)}$ an interference pattern (grating), which contains information about the amplitude and phase relations between the contributing waves. This information is recovered via Bragg scattering using a third beam (backward pump) and gen-

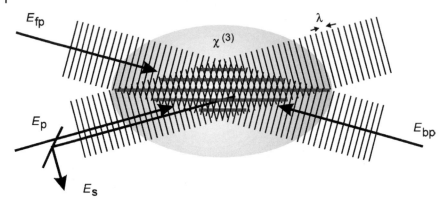

Fig. 6.12 Generation of a holographic grating in a medium with nonlinear susceptibility $\chi^{(3)}$ by interference of a forward propagating incoming light wave (field strength E_{fp}) of wavelength λ, a backward propagating wave (E_{bp}) and an object or probe wave (E_p). The generated signal wave (E_s) is directed into the detector using a beam splitter. Reprinted from Rubahn (1999). Copyright 1999, with permission from John Wiley & Sons Limited.

erates a phase conjugate signal wave. In contrast to conventional holography, in the case of DFWM generation and recovery processes occur simultaneously.

"Phase conjugation" in this context means that the signal wave has the same wavefronts and phase relations as the object wave. Only the sign of the wavevector $\vec{k}$ has changed. For example, the original light wave might have passed through a phase-disturbing medium (e.g., an adsorbate) onto the nonlinear medium (e.g., a surface film). As long as the irradiated signal wave travels through the same phase-disturbing medium, the information content does not get lost.

In the case of a resonant transition involving population transfer, the generated grating might be a density grating. But even without strong absorption the coherent superposition of the laser beams results in a modulation of the complex index of refraction of the medium (amplitude- or polarization-grating), which leads to the generation of a phase-conjugate signal wave. The signal intensity is a measure of the depth of modulation of the grating, just as in the case of diffraction of an external laser beam from the transient grating structure. One might also name the whole process "transient grating scattering" . The lattice constant Λ (and thus the number of lattice rods within the overlap volume of the laser beams, i.e, the sensitivity of the method) depends on the crossing angle Θ via $\Lambda = \frac{\lambda}{2\sin(\Theta/2)}$. Thus a smaller crossing angle results in a more sensitive optical detection.

While the break of inversion symmetry at the surface makes nonlinear optical processes of second order intrinsically surface-sensitive in the case of centro-symmetrical solids, this is no longer the case for a nonlinear process of third order. Hence one has to induce surface-sensitivity externally. This

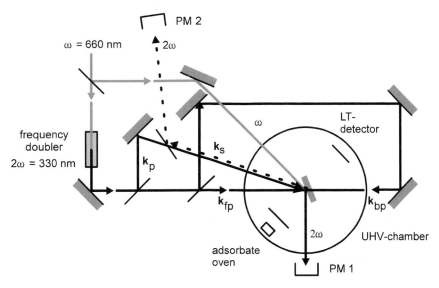

Fig. 6.13 Setup for the simultaneous observation of SHG (using ω-light) and DFWM (using 2ω-light from the same laser source) in the PC geometry from thin cluster films on transparent substrates. The sample is situated in an ultrahigh vacuum chamber and alkali films are deposited from an alkali oven while monitoring the frequency-doubling and the four-wave mixing signals induced by the growing adsorbate film. The deposition rate of adsorbates is calibrated using a Langmuir–Taylor hot-wire detector. PM1 and PM2 are two photomultipliers, which monitor solely 330 nm light. Mirrors are denoted by thick lines and beam splitters by thin lines. Reprinted from Rubahn (1999). Copyright 1999, with permission from John Wiley & Sons Limited.

might be done, for example, by performing nonlinear optics at adsorbates on the surface (Balzer and Rubahn 1995, 1998), such as rough alkali cluster films. A possible setup for an experiment that simultaneously exploits second harmonic generation and four-wave mixing is shown in Fig. 6.13.

The nonlinearly optically active medium is an array of spheroidal alkali clusters with index of refraction n, which gives rise to a DFWM signal intensity (Fisher 1983):

$$I_{\mathrm{DFWM}} = \frac{\mu_0 \omega^2}{\epsilon_0^3 n^4 c^2} l^2 |\chi_{\mathrm{eff}}^{(3)}(\omega_s; \omega_{\mathrm{fp}}, \omega_p, \omega_{\mathrm{bp}})|^2 \times I_{\mathrm{fp}} I_p I_{\mathrm{bp}} \qquad (6.27)$$

To predict the signal intensity one needs a value of the macroscopic susceptibility $\chi^{(3)}$, which is given by the microscopic hyperpolarizability $\alpha^{(3)}$ and the volume of the metallic sphere V:

$$\chi^{(3)} = \alpha^{(3)} / V \qquad (6.28)$$

Again, as in the case of $\chi^{(2)}$ (Eq. (6.24)), the hyperpolarizability will show large values in the case of resonance between the laser frequency ω_{L} and the interband transition frequencies. Numerical values of the electric dipole $\chi^{(3)}$

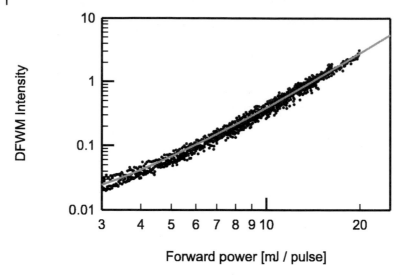

Forward power [mJ / pulse]

Fig. 6.14 DFWM intensity, induced in the PC geometry by 330 nm light, irradiating a distribution of alkali islands ($r_0 = 50$ nm, FWHM 50%) grown on mica at 150 K surface temperature. Dependence on laser power P with a fit curve that resembles a P^3-dependence including an additive linear term which represents reflected background light. Reprinted from Rubahn (1999). Copyright 1999, with permission from John Wiley & Sons Limited.

might be calculated via density matrix theory as exemplified for gold spheres in Hache et al. (1986). For larger particles, the observed nonlinear optical signal is dominated close to electronic resonances by local field enhancement effects. The enhancement can be so strong that one is able to obtain DFWM signal intensities even for irradiation with cw lasers (Balzer and Rubahn 1995). A local field enhancement of a factor of 900 compared with a flat silver surface was found via FWM for an ordered array of equally shaped silver ellipsoids with semi-axes $a = 50$ nm and $b = 150$ nm (Chemla et al. 1983).

In Fig. 6.14 the measured DFWM signal intensity is presented from a cluster film with a mean radius of 50 nm, adsorbed on mica and irradiated by a pulsed laser at 330 nm. The dependence of the signal intensity on the forward laser power is shown with backward and probe power increasing simultaneously (the total value of the backward power was twice that of the forward power). The fitted curve reveals a cubic dependence on laser power, as expected from Eq. (6.27).

Another way to make a four-wave mixing process surface-sensitive is to use the evanescent part of the electromagnetic field at interfaces for spectroscopic purposes. Here one takes advantage of the field enhancement that takes place at the surface of a prism, which is coated by a thin metallic film.

In a four-wave mixing experiment by Shen and coworkers, employing CARS (Chen et al. 1979) a glass prism was coated by a thin silver film (less than 1 μm thick) and surface plasmons were excited in this film by irradiation

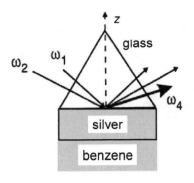

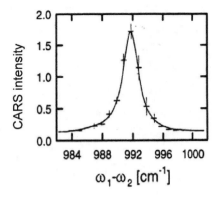

Fig. 6.15 Configuration for the observation of a nonlinear optical signal of third order at a prism surface. The irradiating beams are totally internally reflected and generate anti-Stokes radiation in the benzene, which is strongly enhanced at the spectral position of resonance maxima (right-hand side). Reprinted with permission from Chen et al. (1979). Copyright 1979, American Physical Society.

with visible light. Two surface plasmon waves with wave vectors $\mathbf{k}_1$ and $\mathbf{k}_2$, which travel parallel to the prism surface, were generated by two laser beams ω_1 and ω_2, which were irradiating the prism under the angle of total internal reflection (Fig. 6.15). The irradiances of the two laser beams (2.5 mJ/cm^2 and 25 mJ/cm^2) were low enough to avoid heating effects in the metal.

At the interface between the silver film and the nonlinear medium (benzene in this case) an anti-Stokes wave with wavevector $\mathbf{k}_a = 2\mathbf{k}_1 - \mathbf{k}_2$ is generated, which is coupled out of the prism. Figure 6.11a shows, with the help of an energy level scheme, the generation of the anti-Stokes wave ω_4 via the Raman resonance.

If the frequency difference $\omega_1 - \omega_2$ coincides with a resonance in the benzene, then the nonlinear signal intensity is strongly increasing (Fig. 6.15). In order to observe this effect, one has to fulfil at least two conditions. (a) Phase matching between the wavevector $\mathbf{K}_A$ of the anti-Stokes surface plasmons and the wavevectors of the irradiating lasers, $(\mathbf{K}_A)_\parallel = 2(\mathbf{k}_1)_\parallel - (\mathbf{k}_2)_\parallel$. The index $\parallel$ denotes the components along the surface. The phase-matching condition is fulfilled by choosing the correct angle of incidence of the exciting beams with respect to the surface normal. For the experiment shown, laser beam 1 has to irradiate the prism at an angle of 10°. (b) The frequency of the exciting light has to be in a spectral range where surface plasmons can be excited. Since the plasmon resonances in thin films are spectrally broad (a few tens of nanometers), this condition is easily fulfilled.

Surface CARS is of interest especially for the spectroscopy of materials with strong absorption and fluorescence, since the effective interaction length of the laser beams with the medium is limited by the surface plasmons, which are

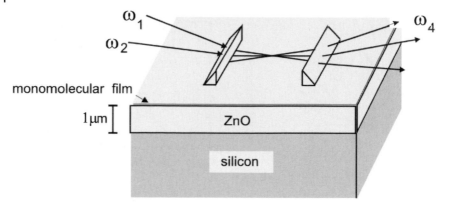

Fig. 6.16 CARS generation in a thin molecular film, adsorbed on a ZnO waveguide. The exciting light frequencies ω_1 and ω_2 are coupled into the waveguide using a prism and the signal ω_4 is coupled out using a second prism. Reprinted from Rubahn (1999). Copyright 1999, with permission from John Wiley & Sons Limited.

excited in the metal, $\lambda_{\parallel}/2\pi = 1/K_{\parallel} \approx 10$ μm. Surface sensitivity is given since the evanescent wave has enough field strength to induce a nonlinear effect of third order only in a layer of thickness $\lambda/6\pi$.

A valuable extension of surface-CARS spectroscopy using a prism is to employ waveguides (Stegemann et al. 1983). The waveguide structure usually consists of a silicon substrate with index of refraction n_1, an oxide film of about 1 μm thickness (index of refraction n_2, e.g., ZnO) and a cover of the nonlinear material to be investigated, which has an index of refraction that is smaller than n_2. For $n_2 > n_1$, an electromagnetic wave that is irradiated into the transparent oxide or polymer film will propagate solely along this material since total internal reflection takes place at the upper and lower sides. More details on the propagation conditions of electromagnetic fields in waveguides and the generated modes can be found, for example, in Yariv (1985). In order to perform CARS spectroscopy at the nonlinear medium, the exciting laser beams are coupled into the waveguide by an attached prism. A few millimeters downwards they are coupled out of the waveguide together with the generated anti-Stokes signal using another prism (Fig. 6.16).

The method allows one to perform nonlinear spectroscopy even with small laser power, since the field strength in the spatially restricted waveguide structure is enhanced and the interaction between nonlinear material and laser beams occurs over a wide spatial range (the CARS intensity increases quadratically with interaction length). In this way, with pulse energies of 0.1 mJ, one might obtain power densities of $200\,\text{MW}/\text{cm}^2$ in a 1 μm thick polystyrene film on silicon. At those power densities along a beam path of 5 mm 5% of the incoming laser, energy is used for the generation of a CARS signal in a benzene stretch vibration (Stegemann et al. 1983).

Simultaneously, a high spectral resolution is possible, which contrasts conventional infrared or electron energy loss spectroscopies. In this way, for example, different physisorption and chemisorption sites of ethylene on ZnO can be identified via their Raman resonances (Wijekoon et al. 1987). The structure and electronic properties of monomolecular organic films can also be investigated sensitively using this technique.

It should be mentioned that all of the above applications of four-wave mixing become especially attractive if one uses short or ultrashort laser pulses, which open up the possibility to study dynamic phenomena in adsorbates on surfaces.

Finally, the application of higher order nonlinear optical processes such as third-harmonic generation (Berkovic 1995; Tsang 1995) or fourth-harmonic generation (Lee et al. 1997) could provide even more detailed interface information. For example, in the case of fourth-harmonic generation the induced polarization is given by $P_j(4\omega) = \sum_{k,l,m,n} \chi^{(4)}_{jklmn} E_k(\omega) E_l(\omega) E_m(\omega) E_n(\omega)$, i.e., the hyperpolarizability is a tensor of rank 5. Hence it is possible to resolve up to five-fold surface symmetries. However, the absolute values of $\chi^{(4)}_{jklmn}$ are small, and thus one needs ultrafast pulses in order to obtain significant signal intensities. The fact that

$$ I_{4\omega} \propto \frac{1}{\tau^3} \tag{6.29} $$

makes the use of femtosecond lasers especially attractive.

6.5
Two-photon Fluorescence Spectroscopy of Adsorbed Atoms and Molecules

In Section 5.4.1 we have considered one-photon fluorescence spectroscopy of adsorbed atoms and molecules. There, the frequency of the emitted light was shifted with respect to that of the incident light allowing one to distinguish emitted light from the signal induced by light scattered by the substrate. Another way to obtain a large signal-to-noise ratio is to use a two-photon scheme of adsorbate excitation.

Alkali metal atoms represent a suitable model system for studying adsorbate fluorescence due to their simple energy structure and prominent radiative properties. Figure 6.17 shows the energy levels of a sodium atom. The ground state $5S_{1/2}$ can be excited either in two steps via successive $3S_{1/2} \rightarrow 3P_{3/2}$ and $3P_{3/2} \rightarrow 5S_{1/2}$ transitions, or via a direct two-photon transition $3S_{1/2} \rightarrow 5S_{1/2}$. The excited atoms then decay via the state $4P_{3/2}$ back to the ground state, accompanied by the emission of UV light. Such an excitation scheme was used for the investigation of fluorescence of Na atoms separated from a metal surface by a spacer layer (Balzer et al. 1993).

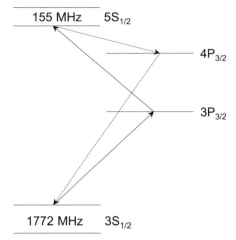

Fig. 6.17 Schematic of the energy levels of a Na atom. The transitions excited by lasers are shown by solid lines, whereas those corresponding to spontaneous decay are shown by dashed lines.

Na atoms were deposited onto the spacer layer from a thermal Na source for a certain amount of time. Thereafter, the source was shut off. The subsequent evolution of Na fluorescence spectra observed on two different substrates is shown in Fig. 6.18. The spectra were obtained by scanning the frequency of one laser around the $3S_{1/2} \rightarrow 3P_{3/2}$ transition frequency, while keeping the frequency of the second laser at the $3P_{3/2} \rightarrow 5S_{1/2}$ transition frequency. The resulting UV light was observed behind an interference filter using an UV-sensitive photomultiplier. The four (barely resolved) peaks in the spectra originate from the large hyperfine splitting of the ground state (see also Fig. 7.13); two of them correspond to a one-photon resonance with the $3S_{1/2} \rightarrow 3P_{3/2}$ transition whereas the other two arise from a two-photon resonance with the $3S_{1/2} \rightarrow 5S_{1/2}$ transition. The hyperfine splittings of the other states are not resolved. In both cases, the fluorescence lines broaden and shift to lower frequencies with time, indicating diffusion of Na atoms toward the spacer layer (i.e., closer to the metal surface).

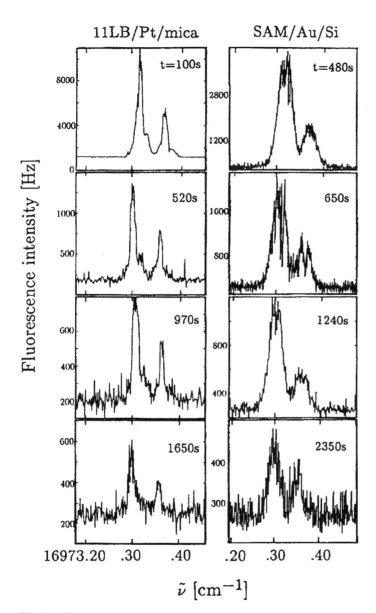

Fig. 6.18 Two-photon fluorescence spectra of Na on top of 11 cadmium-arachidate layers on the Pt surface (left-hand side) and on top of a single dodecanthiol layer on the Au surface (right-hand side) at different times after shutting off the Na source. Reprinted with permission from Balzer et al. (1993). Copyright 1993, American Institute of Physics.

Further Reading

P.N. Butcher and D. Cotter, *The Elements of Nonlinear Optics*, Cambridge University Press, Cambridge, 1990.

T.F. Heinz, Second-Order Nonlinear Optical Effects at Surfaces and Interfaces in *Nonlinear Surface Electromagnetic Phenomena*, H.-E. Ponath, G. I. Stegeman (Eds.), North-Holland, Amsterdam, 1991.

J.F. McGilp, D. Weaire, C.H. Patterson (Eds.), *Epioptics – Linear and Nonlinear Optical Spectroscopy of Surfaces and Interfaces*, Springer-Verlag, Berlin, 1995.

Y.R. Shen, *The Principles of Nonlinear Optics*, John Wiley & Sons, New York, 2002.

Problems

Problem 6.1. A 10 nm thick layer of Si is buried between two 1 μm thick layers of a centrosymmetric medium. Which technique could be used to investigate solely the electronic properties of the buried layer? Why are electron energy loss techniques not applicable?

Problem 6.2. Assume that a metallic surface is irradiated by intense s-polarized light. Would you expect to observe frequency-doubled radiation?

Problem 6.3. Assume that the angle of incidence of the irradiating laser to a thin metallic surface is changed and the generated second harmonic signal is measured in transmission. Would you expect to see a change in signal intensity with changing angle of incidence, and how would that change appear qualitatively?

Problem 6.4. How much does one gain in second harmonic intensity if one irradiates a SHG-active surface with a 100 fs pulse from a laser with energy 1 μJ as compared with an irradiation by a 10 ns pulse of a laser with the same energy ?

Problem 6.5. An electromagnetic wave hits a metal surface and generates a second-order nonlinear optical signal. What is the typical thickness of the surface layer that contributes to the signal and which physical effect is of the main importance in determining this length scale?

7
Optical Spectroscopy at a Gas–Solid Interface

The investigations of optical spectra of gases in the vicinity of solid surfaces date back to Wood in 1909 when he first observed the reflection of light from a glass–mercury vapor interface (Wood 1909). The frequency of light was close to the transition frequency of Hg atoms leading to the name of the method: "selective reflection". The reflection spectrum had a dispersion-like form and its width corresponded to Doppler broadening, in complete agreement with the classical theory of dispersion. However, in 1954, Cojan (1954), who observed selective reflection in the same system but at lower vapor pressure, found evidence for a spectral narrowing when he compared his measured spectra with that calculated including Doppler broadening. Subsequent investigations, both experimental and theoretical, have demonstrated that this phenomenon originates from collisions of the vapor atoms with the glass surface. Their transient behavior after the scattering at the surface gives rise to a *nonlocal* optical response of the gas. This discovery has provided many opportunities for studying gas–surface interactions by pure optical means.

In this chapter we shall consider the optical response of a gas in the close vicinity of a solid surface. We shall become aware of different optical techniques which allow us to study various aspects of interactions between gas atoms or molecules and a surface. They are based on reflection of light from a gas–solid interface. Depending on whether the incidence angle is less or greater than the critical angle, one distinguishes between *selective reflection spectroscopy* (SRS) and *evanescent wave spectroscopy* (EWS).

7.1
Optical Response of a Gas Near a Solid

Let us consider an interface between a solid with refractive index n_1, occupying the half-space $z < 0$, and a gas occupying the half-space $z > 0$. We assume for simplicity that absorption of light in the solid can be neglected and the gas

Optics and Spectroscopy at Surfaces and Interfaces. Vladimir G. Bordo and Horst-Günter Rubahn
Copyright © 2005 WILEY-VCH Verlag GmbH & Co. KGaA, Weinheim
ISBN: 3-527-40560-7

is so rarefied that its refractive index differs only slightly from unity. Let us consider reflection and refraction of a plane electromagnetic wave of the form (3.1) which hits the interface from the side of the solid. Such a problem resembles that discussed in Section 3.1.1. However, in the present case the optical response of the gas (medium 2) cannot be described by a local connection between the induced polarization and the electric field of the wave (Schuurmans 1976).

Reflected field

The electromagnetic field penetrating into the gas induces a polarization of the gas atoms[1] which acts as a source of secondary radiation. It can be written in the form

$$\mathbf{P}(\mathbf{r}, t) = \mathbf{P}_0(z)e^{i\mathbf{k}_{i\|}\cdot\mathbf{r}_\|}e^{-i\omega t} \tag{7.1}$$

where $k_{i\|}$ is given by Eq. (3.5). The contribution of this polarization to the reflected field can be found from the wave equation for the vector potential (Nienhuis et al. 1988). As a result, the reflected field amplitude is represented as a sum of two contributions

$$\mathbf{E}_r = \mathbf{E}_{r0} + \mathbf{E}_{rp} \tag{7.2}$$

where $\mathbf{E}_{r0}$ is given by the Fresnel formulas for a solid–vacuum interface and $\mathbf{E}_{rp}$ originates from the gas polarization.

In the case of s-polarization, both components are perpendicular to the plane of incidence and

$$E_{rp} = \frac{\omega}{c(k_{iz} + k_{tz})} S \tag{7.3}$$

where k_{iz} and k_{tz} are given by Eqs. (3.6) and (3.7), respectively, with $\epsilon_2 = 1$, and

$$S = 4\pi i \frac{\omega}{c} \int_0^\infty P_0(z)e^{ik_{tz}z}dz \tag{7.4}$$

For p-polarized incident waves, the reflected field has the form

$$\mathbf{E}_r = (-E_r \cos\theta_i, 0, -E_r \sin\theta_i) \tag{7.5}$$

In this case the part originating from the gas polarization is given by

$$E_{rp} = \frac{cn_1}{\omega} \frac{k_{tz}^2 - k_{i\|}^2}{n_1^2 k_{tz} + k_{iz}} S \tag{7.6}$$

Gas polarization

In a rarefied gas one can neglect the contribution of the gas polarization to the field which the atoms experience. The polarization can be therefore calculated

---

[1] For definiteness we talk about an atomic vapor. However, the consideration given below can also be applied to molecular gases.

from the amplitude of the transmitted wave which would exist at $z > 0$ in a vacuum. This assumption implies that the reflected field (7.2) is found to first order in the gas density. The assumption is valid as long as the refractive index of the gas deviates negligibly from unity.

The transmitted field at the solid–vacuum interface is described by Fresnel's formulas and can be written as

$$\mathbf{E}_t(\mathbf{r}, t) = \mathbf{E}_{t0} e^{i(\mathbf{k}_{t\|} \cdot \mathbf{r}_\| + k_{tz} z)} e^{-i\omega t} \tag{7.7}$$

with $\mathbf{k}_{t\|} = \mathbf{k}_{i\|}$. For s-polarized incident light

$$E_{t0} = \frac{2k_{iz}}{k_{iz} + k_{tz}} E_{i0} \tag{7.8}$$

whereas for p-polarized light

$$\mathbf{E}_{t0} = \frac{c}{\omega}(-k_{tz}, 0, k_{t\|}) E_{t0} \tag{7.9}$$

with

$$E_{t0} = \frac{n_1 k_{iz}}{n_1^2 k_{tz} + k_{iz}} E_{i0} \tag{7.10}$$

The polarization results from the dipole moments, $\vec{\mu}(\mathbf{r}, t)$, of the gas atoms induced by the electric field (7.7). It can be calculated in the framework of the atomic density matrix formalism. The main features of the phenomena, however, are also reproduced by the classical model of a damped oscillator. We are interested in the case where the frequency of light, ω, is close to the atomic transition frequency, ω_0. The time evolution of the atomic dipole moment can be described, therefore, by the equation for an oscillator driven by the external force $e\mathbf{E}_t(\mathbf{r}, t)$ (Landau and Lifshitz 1978)

$$\frac{d^2\mu}{dt^2} + 2\gamma \frac{d\mu}{dt} + \omega_0^2 \mu = \frac{e^2 f}{m_e} E_{t0} e^{i\mathbf{k}_t \cdot \mathbf{r}} e^{-i\omega t} \tag{7.11}$$

where the damping γ accounts for both the radiative and the collisional relaxation, f is the oscillator strength, e and m_e are the electron charge and mass, respectively, and we have taken into account that $\vec{\mu}$ is parallel to the vector $\mathbf{E}_t$. In the vicinity of the resonance, where $|\omega - \omega_0| \sim \gamma \ll \omega_0$, Eq. (7.11) can be reduced to a first-order equation

$$\frac{d\mu}{dt} + (\gamma + i\omega_0)\mu = \frac{ie^2 f}{2m_e\omega_0} E_{t0} e^{i\mathbf{k}_t \cdot \mathbf{r}} e^{-i\omega t} \tag{7.12}$$

The general solution of Eq. (7.11) can be written in the form

$$\mu(t) = c_- e^{\alpha_- t} + c_+ e^{\alpha_+ t} + c_0 e^{-i\omega t} \tag{7.13}$$

where $\alpha_\pm = -\gamma \pm \sqrt{\gamma^2 - \omega_0^2} \approx -\gamma \pm i\omega_0$ and $c_\pm$ and c_0 are constants. The contribution proportional to $\exp(\alpha_+ t)$ gives rapidly oscillating terms, whereas that proportional to $\exp(\alpha_- t)$ leads to the slow temporal behavior $\sim \exp[-i(\omega - \omega_0)t]$. As a result, the solution can be approximated as

$$\mu(t) \approx \tilde{\mu}(t) e^{\alpha_- t} \tag{7.14}$$

with $\tilde{\mu}(t)$ a slowly varying function of time. Now, substituting (7.14) into Eq. (7.11) and neglecting the second derivative of $\tilde{\mu}(t)$, we arrive at Eq. (7.12).

If an atom moves with a velocity $\mathbf{v}$, the time derivative in Eq. (7.12) should be considered as a total derivative, i.e.,

$$\frac{d}{dt} = \frac{\partial}{\partial t} + \mathbf{v} \cdot \nabla \tag{7.15}$$

Let us make the substitution

$$\mu(\mathbf{r}, t) = \sigma(\mathbf{r}, t) e^{i\mathbf{k}_t \cdot \mathbf{r}} e^{-i\omega t} \tag{7.16}$$

in Eq. (7.12) and let us omit the derivatives $\partial\sigma/\partial t$, $\partial\sigma/\partial x$ and $\partial\sigma/\partial y$ in the steady-state limit ($t \to \infty$). Then we obtain the equation

$$v_z \frac{d\sigma}{dz} + [\gamma - i(\Delta - \mathbf{k}_t \cdot \mathbf{v})]\sigma = \frac{ie^2 f}{2m_e \omega_0} E_{t0} \tag{7.17}$$

where $\Delta = \omega - \omega_0$ is the detuning from resonance. Note that the term $\mathbf{k}_t \cdot \mathbf{v}$ in the brackets accounts for the Doppler shift.

The solution of Eq. (7.17) depends on the boundary conditions which one imposes on $\sigma(z)$ which, in turn, are determined by the processes of excitation and de-excitation of atoms near the surface. We shall consider a situation where the mean free path of the gas atoms exceeds the thickness of the gas boundary layer where the optical response is formed. Accordingly, separate boundary conditions are set for atoms moving to the surface and for those departing from it. The most simple, but often fairly realistic, assumption is that all atoms arriving at the surface are adsorbed on it and then are desorbed with completely quenched polarization, i.e.,

$$\sigma_+(z = 0) = 0 \tag{7.18}$$

For atoms moving to the surface it is reasonable to assume that their polarization is vanishing at infinite distances, i.e.,

$$\sigma_-(z \to \infty) = 0 \tag{7.19}$$

Here and in the following the subscripts "+" and "−" denote the contributions of atoms having positive and negative v_z, respectively.

The corresponding solutions of Eq. (7.17) give the induced dipole moment of atoms arriving at the surface

$$\mu_-(\mathbf{r}, t; \mathbf{v}) = \alpha(\omega; \mathbf{v}) E_{t0} e^{i\mathbf{k}_t \cdot \mathbf{r}} e^{-i\omega t} \tag{7.20}$$

and that of the scattered atoms

$$\mu_+(\mathbf{r}, t; \mathbf{v}) = \alpha(\omega; \mathbf{v}) \left\{ 1 - \exp\left[-\frac{\gamma - i(\Delta - \mathbf{k}_t \cdot \mathbf{v})}{v_z} z \right] \right\} E_{t0} e^{i\mathbf{k}_t \cdot \mathbf{r}} e^{-i\omega t} \tag{7.21}$$

with

$$\alpha(\omega; \mathbf{v}) = \frac{ie^2 f}{2m_e \omega_0} \frac{1}{\gamma - i(\Delta - \mathbf{k}_t \cdot \mathbf{v})} \tag{7.22}$$

denoting the atomic polarizability. Note that the exponential term in the curly brackets, Eq. (7.21), describes the transient behavior of the induced dipole moments of atoms moving from the surface. It vanishes at distances $z \gg v_z/\gamma$.

The gas polarization, Eq. (7.1), is found as the dipole moment averaged over atomic velocities and is given by

$$\mathbf{P}_0(z) = \chi(z) \mathbf{E}_{t0} e^{ik_{tz} z} \tag{7.23}$$

where

$$\chi(z) = \chi_- + \chi_+(z) \tag{7.24}$$

with

$$\chi_- = n \int_{v_z \leq 0} f_M(\mathbf{v}) \alpha(\omega; \mathbf{v}) d\mathbf{v} \tag{7.25}$$

and

$$\chi_+(z) = n \int_{v_z > 0} f_M(\mathbf{v}) \alpha(\omega; \mathbf{v}) \left\{ 1 - \exp\left[-\frac{\gamma - i(\Delta - \mathbf{k}_t \cdot \mathbf{v})}{v_z} z \right] \right\} d\mathbf{v} \tag{7.26}$$

Here, n is the number density of the gas atoms and f_M is the Maxwellian velocity distribution function, Eq. (2.152). We notice that the spatial dependence of the contribution from the arriving atoms follows the spatial variation of the external field, whereas that is not the case for the contribution of the scattered atoms because of the exponential term in the curly brackets. In other words, the optical response of the gas near the surface is *nonlocal*.

Reflectivity

From a combination of Eqs (7.2), (7.4) and (7.23) the reflectivity $R = |E_r/E_{i0}|^2$ is calculated. To first order in the number density of atoms, the reflectivity can be written in the form

$$\frac{\Delta R_s}{R_s^0} = -16\pi \frac{\omega}{c} \frac{n_1 \cos \theta_i}{n_1^2 - 1} \operatorname{Im}\left[\int_0^\infty \chi(z) e^{2ik_{tz} z} dz \right] \tag{7.27}$$

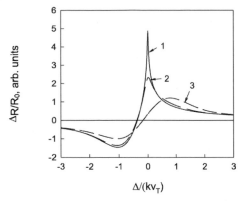

Fig. 7.1 The frequency dependence of the reflectivity at normal incidence in the vicinity of the resonance $\omega = \omega_0$ for: $\gamma/kv_T = 0.01$ (1); $\gamma/kv_T = 0.1$ (2); $\gamma/kv_T = 1$ (3).

for s-polarization, and

$$\frac{\Delta R_p}{R_p^0} = -16\pi\frac{\omega}{c}\frac{n_1\cos\theta_i(1 - 2n_1^2\sin^2\theta_i)}{(n_1^2 - 1)[1 - (n_1^2 + 1)\sin^2\theta_i]}\,\mathrm{Im}\left[\int_0^\infty \chi(z)e^{2ik_{tz}z}dz\right] \quad (7.28)$$

for p-polarization. Here R^0 is the reflectivity at the solid–vacuum interface and $\Delta R = R - R^0$. Note that at normal incidence ($\theta_i = 0$) these expressions coincide with each other.

Figure 7.1 shows the frequency behavior of the relative reflectivity near the atomic transition frequency for different ratios between damping γ and Doppler width kv_T. It should be noted that, if the mean length scale at which the transient term in $\chi_+(z)$ decays is much less than that of the external field variation along the normal to the surface, i.e.,[2]

$$\frac{v_T}{\gamma} \ll \frac{\lambda}{2\pi\cos\theta_t} \quad (7.29)$$

then the exponential term gives a negligible contribution to the integration of $P_0(z)$ in Eq. (7.4). In this limit the function $\chi(z)$ is reduced to the linear susceptibility of the gas

$$\chi(\omega) = n\int f_M(\mathbf{v})\alpha(\omega;\mathbf{v})d\mathbf{v} \quad (7.30)$$

whose frequency dependence is described by a Voigt-type lineshape. This conclusion follows also from conventional dispersion theory which assumes a local relationship between the induced polarization and the external field

---

2) Here, v_T is the most probable thermal velocity in the gas, Eq. (2.153).

(Born and Wolf 1975). For normal incidence, the condition (7.29) can be rewritten as

$$\gamma \gg k_t v_T \tag{7.31}$$

which means that the homogeneous linewidth significantly exceeds the Doppler broadening.

In the case of total internal reflection, where k_{tz} is pure imaginary and the transmitted wave is evanescent, the results obtained above are still valid if one sets $k_{tz} = i\kappa$ with κ given by Eq. (3.38) for $n_2 = 1$.

7.2
Selective Reflection Spectroscopy

As we have seen in the preceding section, the reflectivity at the gas–solid interface contains the contribution from atoms with $v_z \leq 0$ determined by χ_-, Eq. (7.25). At normal incidence the vector $\mathbf{k}_t$ is parallel to the z-axis and the polarizability $\alpha(\omega; \mathbf{v})$ depends only on the v_z component. Accordingly,

$$\chi_- = \frac{n}{\sqrt{\pi} v_T} \frac{ie^2 f}{2m_e \omega_0} \int_{-\infty}^{0} \frac{\exp(-v_z^2/v_T^2)dv_z}{\gamma - i(\Delta - k_t v_z)} \tag{7.32}$$

Near the resonance ($\Delta = 0$) in the limit of negligible γ, the asymptotic expansion of the integral in Eq. (7.32) contains the leading term $\sim \ln|\gamma - i\Delta|$ which means that the reflectivity line has a sub-Doppler feature determined by the linewidth γ (see Fig. 7.1). This feature originates from the contribution of atoms which have $v_z = 0$, i.e., which move parallel to the surface, allowing one to measure the homogeneous broadening given by γ on the background of a much wider Doppler broadening. In addition, the optical response of such atoms contains information on the frequency shift near the surface given by the dependence $\omega_0(z)$. In this section we shall consider various applications of selective reflection (SR) to spectroscopy of atomic vapors in the vicinity of a surface.

7.2.1
Spectral Narrowing of Selective Reflection

Although Cojan was the first to observe spectral narrowing in SR, systematic studies started much later (Woerdman and Schuurmans 1975). The reflectivity at a glass–sodium vapor interface near the sodium D_1 and D_2 lines was measured for different vapor densities. At low sodium densities the resolution of the hyperfine doublet splitting clearly indicated a considerable spectral narrowing as compared to the Doppler width. With an increasing sodium density the hyperfine structure of the SR spectrum was smeared out. When the

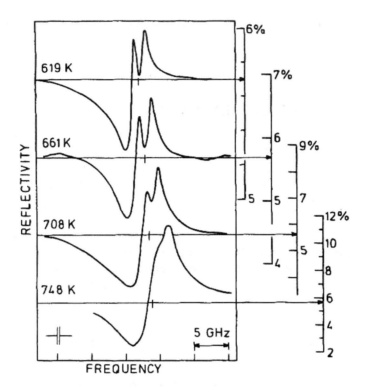

Fig. 7.2 Reflectivity from a glass–sodium vapor interface at normal incidence for different sodium temperatures. Higher temperatures correspond to higher sodium vapor densities. Reprinted from Woerdman and Schuurmans (1975), Copyright 1975, with permission from Elsevier.

homogeneous linewidth was comparable with the Doppler broadening, the splitting almost disappeared (Fig. 7.2).

This behavior is consistent with the theoretical consideration given in the preceding section. If the vapor density is increasing, the homogeneous linewidth γ is also increasing due to interatomic collisions in the vapor. For $\gamma \geq k_t v_T$ the effect of nonlocality in the optical response becomes insignificant and the SR lineshape tends to a Voigt profile determined by the Doppler width.

The spectral narrowing in SR can be used for investigations of both the collisional broadening and the shift at the center of the absorption line of the vapor (Sautenkov et al. 1981; Akul'shin et al. 1982). Both processes are difficult to measure by conventional methods because of strong absorption. SR can be applied also to the study of coherence between Zeeman sublevels (Weis et al. 1992). However, if a gas cannot be considered as rarefied, it is necessary to take into account the effect of a local field (Guo et al. 1994). Also, it should be pointed out that the theory of SR must be modified to account for power broadening if the incident light saturates the atomic transitions (Vartanyan 1985).

7.2.2
Observation of Transient Effects

In Section 7.1 we have seen that the optical response of atoms scattered by the surface contains a transient contribution which is essential at distances $z \leq v_T/\gamma$ (see Eq. (7.26)). If the gas volume which contributes to the optical signal has a thickness much larger than v_T/γ, the transient term can be neglected and the optical spectrum is reproduced by the conventional theory of dispersion. One way to study the transient polarization behavior is to observe the gas fluorescence excited by normally incident light at an angle larger than the critical angle (Burgmans et al. 1977).

An excited atom located at the point $\mathbf{r}_a$ can be considered as an oscillating dipole $\vec{\mu}(\mathbf{r}_a, t; \mathbf{v})$ given by Eqs (7.20) or (7.21) depending on the sign of v_z. Its radiation field is given by a two-dimensional Fourier integral

$$\mathbf{E}(\mathbf{r}, t) = -\frac{i}{2\pi} \int \frac{d\mathbf{k}_{\|}}{k_z} e^{-i\mathbf{k}_{\|} \cdot \mathbf{R}_{\|}} e^{-ik_z R_z} [(\vec{\mu} \cdot \mathbf{k})\mathbf{k} - k^2 \vec{\mu}] \tag{7.33}$$

where $0 < z < z_a$, $\mathbf{R} = (\mathbf{R}_{\|}, R_z) = \mathbf{r} - \mathbf{r}_a$,[3] and $\mathbf{k} = (\mathbf{k}_{\|}, k_z)$ with $k_z = [(\omega/c)^2 - k_{\|}^2]^{1/2}$. The field (7.33) can be regarded as a superposition of propagating waves with $k_{\|} < \omega/c$ and evanescent waves with $k_{\|} > \omega/c$ which decay at large $|R_z|$. The transmission of each component across the interface is described by the Fresnel formulae. The waves propagating at an angle $\theta_k = \arcsin(k_{\|}c/\omega)$ generate waves in the substrate which can be observed at an angle of detection, θ, satisfying Snell's law:

$$\theta = \arcsin\left(\frac{1}{n_1} \sin \theta_k\right) = \arcsin\left(\frac{k_{\|}c}{n_1\omega}\right) \tag{7.34}$$

Obviously, for such waves we have $\theta < \theta_c$ with θ_c the critical angle. The evanescent waves transmitted across the interface can be observed at $\theta > \theta_c$ (Fig. 7.3). Due to the decaying character of the evanescent field, only atoms at distances

$$z \leq \frac{1}{\text{Im}(k_z)} = \frac{\lambda}{2\pi\sqrt{n_1^2 \sin^2 \theta - 1}} \tag{7.35}$$

contribute to this signal.

Figure 7.4 shows a set of fluorescence excitation spectra taken near the D_1 line of sodium vapor bordering a glass surface. The vapor was excited by

3) Note that $R_z < 0$.

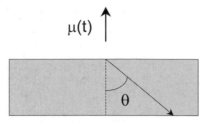

$\mu(t)$ ↑

Fig. 7.3 Radiating dipole near an interface. Its radiation is observed from the glass side at an angle θ which exceeds the critical angle.

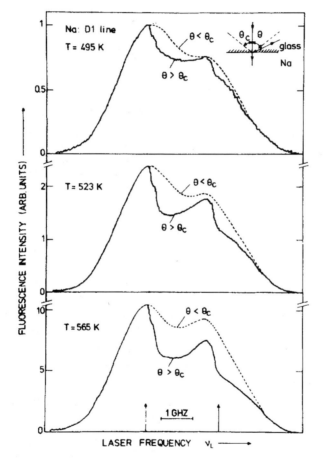

Fig. 7.4 Fluorescence excitation spectra of the D_1 line of sodium at different vapor densities. For comparison, the amplitudes of the spectra observed at $\theta < \theta_c$ are reduced by a factor of about 20. The arrows denote the Na hyperfine structure components. Reprinted with permission from Burgmans et al. (1977). Copyright 1977, American Physical Society.

a laser beam propagating normal to the interface, whereas the fluorescence was detected either at an angle 3° larger than θ_c or at 3° smaller than θ_c. For $\theta < \theta_c$ the spectrum consists of two Doppler-broadened peaks arising from the hyperfine splitting of the ground state (1772 MHz). For $\theta > \theta_c$ both peaks clearly display a cutoff on the high-frequency side.

These results can be understood as follows. In the absence of phase correlations between different atoms, the intensity of the radiation field is determined by the squared modulus of (7.33) averaged over the atomic velocities, i.e.,

$$I(z) \propto \int d\mathbf{v} f_M(\mathbf{v}) |\vec{\mu}(z, t; \mathbf{v})|^2 \tag{7.36}$$

The fluorescence signal observed below the critical angle originates from a large vapor volume of a thickness given by the absorption length, L. The contribution of the transient term in Eq. (7.21) which is essential at $z \leq v_T/\gamma \ll L$ can be neglected. The spectrum is then determined as

$$I(\Delta) \propto \int \frac{f_M(\mathbf{v}) d\mathbf{v}}{(\Delta - \mathbf{k}_t \cdot \mathbf{v})^2 + \gamma^2} \tag{7.37}$$

i.e., is represented by a usual Voigt lineshape.

The fluorescence registered above the critical angle results from radiation of atoms located at $z \leq v_T/\gamma$. At negative detunings from resonance (the laser frequency is less than the atomic transition frequency) only the atoms moving to the interface are excited due to the Doppler effect. They are in the laser field for a long time ($\sim L/v_T$) and their excited states are highly populated. At positive detunings, only the atoms departing from the surface interact effectively with the laser beam. These atoms, which have been de-excited at the surface, do not have enough time to establish steady-state populations. As a result, the right wing of the excitation spectrum is suppressed as compared with the left wing.

The study of the transient behavior of gas atoms near to the surface can be regarded as a general approach which allows one to investigate various relaxation processes in gas–surface scattering. For example, if gas atoms are spin-polarized, e.g., by means of optical pumping, their probing with the use of SR spectroscopy makes it possible to determine the probability of depolarization in atom–surface collisions (Grafström and Suter 1996).

7.2.3
Manifestation of Atom–Surface Interactions

The sub-Doppler structure of the SR spectrum provides an opportunity to determine the frequency shift caused by the atom–surface interaction with high resolution. This singularity can be isolated if one applies a low-frequency modulation to the incident light. By detecting the amplitude-modulated reflected signal one obtains for the reflectivity at the modulation frequency, ω_m,

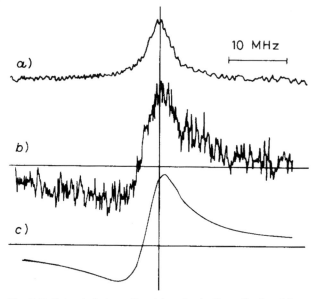

Fig. 7.5 Saturated absorption (a) and selective reflection (b) spectra across the $F = 4 \rightarrow F' = 5$ transition of the Cs D_2 line. (c) Theoretical SR lineshape for $Ck_t^3/\gamma = 0.2$. Reprinted with permission from Oria et al. (1991). Copyright 1991, EDP Sciences.

(Ducloy and Fichet 1991)

$$R(\omega_m) \propto \mathrm{Re}\left(\frac{d\chi_-}{d\omega}\right) = \frac{n}{\sqrt{\pi}v_T}\frac{e^2 f}{2m_e\omega_0 k_t}\frac{\Delta}{\Delta^2 + \gamma^2} \qquad (7.38)$$

i.e., a Doppler-free dispersion lineshape.

This technique was applied to the study of the reflection spectrum at the interface between a glass window and Cs vapor (Oria et al. 1991). The incident light was scanned across the $6S_{1/2}(F = 4) \rightarrow 6P_{3/2}(F' = 3, 4, 5)$ transitions of the caesium D_2 line. The reflection spectrum was compared with the saturated absorption spectrum and a red shift of about 3 MHz, along with a remarkable asymmetry, were observed (Fig. 7.5).

In order to fit the SR spectrum one has to take into account the z-dependence of the atomic transition frequency due to the atom–surface interaction (Section 2.2.1). It can be approximated as

$$\omega_0(z) = \omega_0 - \frac{C}{z^3} \qquad (7.39)$$

with ω_0 the transition frequency of a free atom and C the constant determined by the difference between the interaction potentials in the atomic states involved in the transition. Accordingly, the quantity χ_- which determines the spectral feature at $\Delta \approx 0$ becomes a function of z. Note that the decrease in the

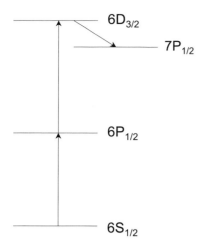

Fig. 7.6 Excitation scheme of Cs atoms.

transition frequency when approaching the surface, which occurs for $C > 0$, leads to a red shift of the SR spectrum relative to ω_0. Figure 7.5c shows the best fit of the SR spectrum obtained for $Ck_t^3/\gamma = 0.2$ ($C \approx 2\,\text{kHz·}\mu\text{m}^3$).

If an atom is in an excited state, the character of the atom–surface potential (attractive versus repulsive) is determined by the sum of the van der Waals interaction (Eq. (2.100)) and the classical shift (Eq. (2.101)). The first term may be either positive or negative. The second term also may have either sign if one of the transition frequencies is close to the frequency of a surface excitation determined by the equation $n_1^2(\omega) + 1 = 0$. As a result, the interaction constant C may be either positive or negative.

The latter situation was observed in SR from the interface between a sapphire surface and Cs vapor (Failache et al. 1999). The SR spectrum was registered at the $6P_{1/2} \rightarrow 6D_{3/2}$ transition while the transition $6S_{1/2} \rightarrow 6P_{1/2}$ was pumped by another laser (Fig. 7.6). An important feature of this gas–solid system is that the wavelength of the transition $7P_{1/2} \rightarrow 6D_{3/2}$ (12.15 μm) is close to the wavelength of surface optical phonons (~ 12 μm) which leads to a large positive contribution to the energy of the $6D_{3/2}$ state. As a consequence, the atom–surface interaction in this state is repulsive (Fichet et al. 1995) and the SR spectrum at the transition $6P_{1/2} \rightarrow 6D_{3/2}$ is blue-shifted and can be fitted with $C \approx -160\,\text{kHz·}\mu\text{m}^3$.

7.2.4
Selective Reflection in a Pump–Probe Scheme

The possibilities of SR spectroscopy can be extended if the reflectivity of a probe (signal) beam is measured while optically pumping the gaseous

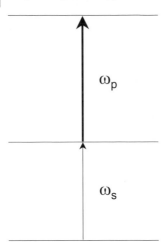

Fig. 7.7 Scheme of cascade transitions among the atomic levels.

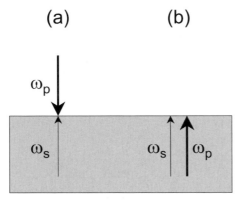

Fig. 7.8 Two geometries of excitation: (a) counterpropagating; (b) copropagating.

medium by another light beam (Rabi et al. 1994). Let us consider the case where both beams are incident normal to the interface, the pump beam (frequency ω_p) is tuned to the upper transition (frequency ω_2) in a three-level cascade scheme, whereas the probe beam (frequency ω_s) is scanned across the lower transition (frequency ω_1) (Fig. 7.7). The two possible excitation configurations are: (a) a counterpropagating geometry with the probe beam incident at the interface from the dielectric and the pump beam incident from the vapor; and (b) a co-propagating geometry with both beams incident from the dielectric side (Fig. 7.8).

Besides the sub-Doppler feature at $\omega_s \approx \omega_1$, which is known from one-photon SR, the two-photon excitation introduces an additional Doppler-free resonance. In the counterpropagating geometry, a sub-Doppler signal is generated when the pump wave detuning is positive, i.e., $\Delta_p = \omega_p - \omega_1 > 0$. Such a wave excites atoms moving to the surface and having $v_z = \Delta_p/k_p$ with $k_p < 0$ the wave vector of the pump wave. These atoms are also in resonance with the probe field if $\Delta_s = \omega_s - \omega_1 = k_s(\Delta_p/k_p)$, with k_s the wave vector of the probe wave. This gives rise to an enhancement of the reflectivity at this ω_s. In a copropagating geometry, a similar reflection resonance also occurs at positive pump wave detuning, but here we have $k_p > 0$ and thus the resonance conditions are fulfilled for atoms departing from the surface.

Figure 7.9 shows the amplitude of the two-photon resonance in SR from the glass/Cs vapor interface plotted as a function of Δ_p for two different cascade excitations of Cs atoms. The left-hand side of each graph corresponds to the atoms arriving at the surface, whereas the right-hand side originates from the atoms leaving it. The signal amplitude at a given Δ_p is proportional to the fraction of atoms having $v_z = \Delta_p/k_p$ and therefore such a plot provides information on the velocity distribution function of both atoms moving to the surface and departing from it. The solid lines in Fig. 7.9 represent the best fits obtained with a Maxwellian velocity distribution for both groups of atoms. At the centers of the plots, which correspond to atoms moving almost parallel to the surface, one can see a remarkable deviation from these fits, possibly to be attributed to the effect of the attractive van der Waals potential on low-velocity atoms.

7.3
Evanescent Wave Fluorescence Spectroscopy

The gas volume probed in SR spectroscopy is determined by the absorption length. For a weakly absorbing gas this quantity can be so large that the contribution of the boundary effects becomes negligible. In addition, in this case the presence of the gas only slightly influences the reflectivity at the interface. The sensitivity to the gas boundary properties can be, however, enhanced in total internal reflection (Section 3.1.2) where an evanescent wave (EW) is excited at the gas–solid interface. In the visible spectral range, its penetration depth into the gas spans a scale of only a few hundred nanometers. This advantage of evanescent wave spectroscopy is widely used in various sensors for detection of gas molecules. EWs in such sensors are implemented via excitation of waveguide modes in planar waveguides or in optical fibers. If the EW frequency is resonant to the optical transition frequency of molecules in the medium bordering the waveguide surface, the fluorescence signal from the molecules or the dielectric losses at the resonant frequencies can be used to obtain the concentration of molecules at the interface.

However, if one investigates not only peak line intensities in EW spectroscopy but also the *lineshapes*, much more detailed physical information can

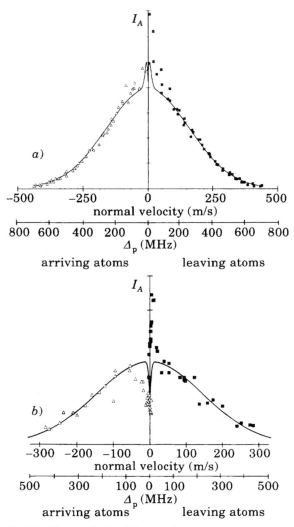

Fig. 7.9 The amplitude of the two-photon resonance as a function of the pump detuning in reflection from the glass/Cs vapor interface; (a) the $6S \rightarrow 6P \rightarrow 8D$ transition; (b) the $6S \rightarrow 6P \rightarrow 9S$ transition. Reprinted with permission from Rabi et al. (1994). Copyright 1994, EDP Sciences.

be obtained. The same advantage can be achieved, in principle, through the excitation of surface polaritons on a crystal surface due to the rapid decay of their amplitude into the gas. In this section we shall consider how evanescent wave fluorescence spectroscopy can be utilized for studying the processes of adsorption, desorption and atom–surface scattering.

7.3.1
One-photon Excitation

In Section 7.1 we have seen that the dipole moments of atoms arriving at the surface, Eq. (7.20), and of those scattered by the surface, Eq. (7.21), have different spatial dependencies. Taking into account that $\mathbf{r}(t) = \mathbf{r}(0) + \mathbf{v}t$, we conclude that the former vary with time as

$$\mu_-(t) \propto \exp[-i(\omega - \mathbf{k}_t \cdot \mathbf{v})t] \qquad (7.40)$$

whereas the latter, besides a similar contribution, also contain a transient term, μ_{tr}, which varies with time as

$$\mu_{tr}(t) \propto \exp[-i(\omega_0 - i\gamma)t] \qquad (7.41)$$

Equation (7.40) implies that the atomic dipoles moving to the surface oscillate with the frequency of the incident light corrected by a Doppler shift. The temporal variation of the atomic dipoles scattered by the surface has two contributions: one of them is also described by Eq. (7.40), whereas the other one corresponds to damped vibrations with the atomic eigenfrequency ω_0. In the classical description, an oscillating dipole emits radiation having the frequency of its oscillations. The same result is obtained in the framework of a rigorous quantum-mechanical approach. Therefore, one expects two lines in the fluorescence spectrum of the gas. One of them originates from both atoms arriving at the surface and departing from it and is centered at the frequency of the incident light. The other one arises only from the atoms desorbed from the surface and its frequency is given by the atomic transition frequency. Due to the Doppler effect in emission for moving atoms we expect Doppler broadening of both lines.

Excitation of the gas by an evanescent wave introduces a pure imaginary z-component of the wave vector $\mathbf{k}_t$ and, hence, an additional time-of-flight broadening of the fluorescence lines. The fluorescence line intensity is determined by the gas volume where the polarization corresponding to the contributions given by either Eq. (7.40) or Eq. (7.41) is essentially nonzero. Therefore, the intensity of emission at the frequency ω is proportional to the EW penetration depth, δ, whereas that at the frequency ω_0 is proportional to the polarization memory length, $l_2 = v_T/\gamma$ (see Section 2.4.3). The latter line is thus dominant in the spectrum if $\delta \ll l_2$.

These conclusions were confirmed experimentally in fluorescence spectra of sodium atoms excited by an EW nearly resonant to the $3S_{1/2} \rightarrow 3P_{3/2}$ transition (Bordo et al. 1997). A flux of sodium atoms was directed onto a glass prism surface mounted in a vacuum chamber. The fluorescence excited by fixed-frequency laser radiation was spectrally resolved with the help of a Fabry–Perot etalon. The observed spectrum is shown in Fig. 7.10. Here, the

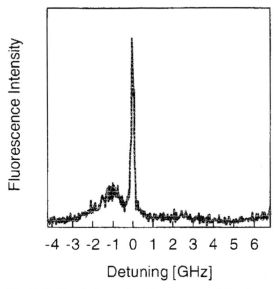

Fig. 7.10 Spectrally resolved fluorescence of Na atoms excited in the evanescent wave at a glass prism surface. Reprinted from Bordo et al. (1997), Copyright 1997, with permission from Elsevier.

narrow line at the frequency of the incident light originates from EW scattering at surface roughness, whereas the broader line represents the fluorescence of Na atoms. The width of this latter line is proportional to the prism temperature and does not depend on the temperature of the incident atomic flux. Under the given experimental conditions, the ratio $v_T/(\gamma\delta)$ was about 67, implying that the desorbed atoms dominate the spectrum. This agrees with the temperature dependence of the fluorescence linewidth. The method thus allows one to study the desorption process into the gas atmosphere.

7.3.2
Two-photon Excitation

Two-photon fluorescence spectroscopy provides a much better signal-to-noise ratio compared with the one-photon excitation scheme. We have already seen its advantages in the course of measuring the fluorescence of atoms adsorbed near to a metal surface (Section 6.5). In this chapter we shall consider a similar technique based on two-photon evanescent wave excitation of a gas near a surface. Due to the reliability of the signal, this kind of spectroscopy allows one to study the fluorescence lineshapes in detail.

In gases, the two-photon excitation has an additional advantage when being applied in the geometry of counterpropagating waves. In this case, the overall Doppler broadening is determined by the quantity $v_T|\omega_1 - \omega_2|/c$. If

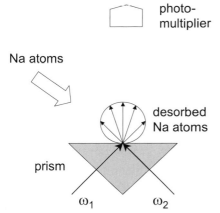

Fig. 7.11 Schematic drawing of the experiment for the study of two-photon evanescent-wave fluorescence from sodium atoms.

the wave frequencies, ω_1 and ω_2, are close to each other, the residual Doppler broadening may be negligible. Then at low gas pressures where γ is small, the broadening of fluorescence lines is dictated mainly by the relaxation of the excitation, while the excited atoms leave the gas volume being "illuminated" by the evanescent waves. The corresponding relaxation times are given by the flight times across the EW fields, δ_1/v_z and δ_2/v_z with δ_1 and δ_2 the EW penetration depths and v_z the atom velocity component along the surface normal. As a result, the fluorescence lineshape is very sensitive to the velocity distribution function along the v_z-component.

This was demonstrated in an experiment on two-photon fluorescence of sodium atoms excited by counterpropagating EWs at a glass prism surface (Bordo and Rubahn 1999b). The prism was mounted inside a vacuum chamber. Sodium atoms from the dispenser were directed in the plane of incidence of the laser beams towards the prism surface (Fig. 7.11). One of the lasers exciting EWs was set in resonance with the $3P_{3/2} \rightarrow 5S_{1/2}$ transition whereas the second laser was scanned across the adjacent $3S_{1/2} \rightarrow 3P_{3/2}$ transition (see Fig. 6.17). The resulting fluorescence light from the $4P_{1/2,3/2} \rightarrow 3S_{1/2}$ transitions was observed as a function of detuning of the second laser. Due to the Doppler shift, the exciting EWs were out of resonance with the atomic flux arriving at the surface and thus the contribution to the fluorescence spectrum was negligible. The same is true for the atoms specularly scattered by the surface. The observed spectrum, therefore, originated mainly from sodium atoms desorbed from the surface.

The intensity of fluorescence induced by the desorbing flux is determined by the quantity N_e/τ with N_e the total number of desorbed atoms in the excited state and τ their radiative decay lifetime. In a steady-state excitation

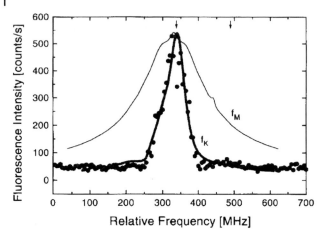

Fig. 7.12 Pair of the hyperfine components (denoted by arrows on top) in the two-evanescent-wave fluorescence spectrum observed from Na atoms desorbing from a glass surface. The lines represent the theoretical spectra calculated with the velocity distribution functions f_K (thick solid line) and f_M (thin solid line). Reprinted with permission from Bordo and Rubahn (1999b). Copyright 1999, American Physical Society.

regime the number of excited atoms can be found as

$$N_e = J \int_{v_z > 0} \int_0^\infty n_e(\mathbf{v}, t) f(\mathbf{v}) d\mathbf{v} dt \tag{7.42}$$

where J and $f(\mathbf{v})$ are the total desorbing flux and its velocity distribution function, respectively, and n_e is the population of the excited state. Here, the time t in the integrand is counted beginning with the moment of desorption. As we have seen in Section 2.4.1, Knudsen's cosine law suggests that the function $f(\mathbf{v})$ has the form of Eq. (2.157). Figure 7.12 represents the two-evanescent-wave fluorescence spectrum obtained by tuning the frequency of the second laser across the $3S_{1/2}(F = 1) \rightarrow 3P_{3/2}(F')$ transitions. Spectra calculated with different velocity distribution functions are also shown. The best fit is obtained with the function f_K, Eq. (2.157), assuming the cosine law for desorption, whereas the curve calculated with the function f_M, Eq. (2.152), implying isotropic desorption, clearly does not describe the observed lineshape.

7.3.3
Temperature Dependence of Fluorescence Spectra

Gas spectra excited by EWs can also be influenced by atoms which are *adsorbed* on the surface. The adsorbate overlayer determines the boundary conditions for Maxwell's equations at the interface and, therefore, also the EW amplitude in the gas. As a result, the gas optical spectra which depend on the EW intensity, contain indirectly information about the adsorbed layer. In particular, the atoms adsorbed on the surface are polarized by the exciting field, $\mathbf{E}_{i0}e^{-i\omega t}$, and thus create a two-dimensional displacement current in the surface plane

$$\mathbf{j}(t) = \frac{\partial \mathbf{P}_s}{\partial t} \tag{7.43}$$

For small surface coverages, when the effect of the local field can be neglected, the adsorbate polarization, $\mathbf{P}_s$, is expressed in terms of the adatom dynamic polarizability, $\alpha_s(\omega)$, as

$$\mathbf{P}_s(t) = N_s \alpha_s(\omega) \mathbf{E}_{i0\|} e^{-i\omega t} \tag{7.44}$$

Here, $\mathbf{E}_{i0\|}$ is the tangential component of the electric field amplitude at the interface and N_s is the surface number density of adsorbed atoms. It is also assumed that the dielectric response of the adsorbed atoms is isotropic in the surface plane. Accounting for the surface current (7.43) in the boundary condition for the tangential components of the magnetic field, results in a modified expression for the transmitted field amplitude. For example, in the case of s-polarization one obtains, instead of Eq. (3.13) with $\epsilon_2 = 1$,

$$E_{t0} = \frac{2\sqrt{\epsilon_1} \cos \theta_i}{\sqrt{\epsilon_1} \cos \theta_i + \sqrt{1 - \epsilon_1 \sin^2 \theta_i + 4\pi i(\omega/c)^2 N_s \alpha_s}} E_{i0} \tag{7.45}$$

This equation is valid also for the case of total internal reflection provided that θ_i exceeds the critical angle. It relates the EW amplitude to the surface coverage, θ, defined in terms of N_s. In the framework of the Langmuir model of adsorption the quantity θ is related to the surface temperature by Eqs (2.146), (2.144) and (2.145). Therefore, measuring the EW fluorescence spectra from the gas as a function of surface temperature allows one to obtain information on the adsorbate.

An experimental realization of this idea was demonstrated for adsorption of sodium atoms at a glass prism surface in the setup described in the preceding section (see Fig. 7.11) (Bordo and Rubahn 1999a). Both EWs were excited near the critical angle in order to increase the EW penetration depths and thus to decrease the time-of-flight broadening. In this regime, the lines in the two-photon fluorescence spectrum split into two if one of the EWs saturates the lower transition (Autler–Townes splitting). The value of the line splitting is proportional to the amplitude of this EW (Fig. 7.13). Figure 7.14 demonstrates that the splitting increases at fixed EW intensity as one increases the temperature of the prism surface while continuously evaporating Na atoms from the dispenser. At relatively high surface temperatures ($T > 240$ K) the splitting reaches a steady-state value indicating that the equilibrium surface coverage tends to zero. Comparison of the theoretical curves with the measured fluorescence spectra allows one to deduce the EW amplitude corresponding to a given surface temperature. Then, using Eq. (7.45), one can calculate the EW

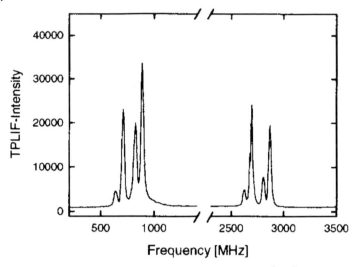

Fig. 7.13 Two-evanescent-wave fluorescence spectra of sodium atoms excited near the critical angles. Each line exhibits Autler–Townes splitting. Reprinted with permission from Bordo and Rubahn (1999a). Copyright 1999, Optical Society of America.

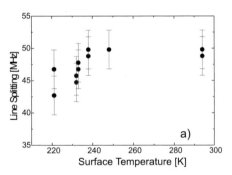
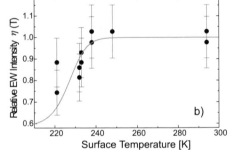

Fig. 7.14 (a) Dependence of the measured Autler–Townes splitting on the prism temperature for fixed EW intensity. (b) Fit to the experimental data, resulting in the EW amplitude vs. surface temperature. With permission from Bordo and Rubahn (1999a). Copyright 1999, Optical Society of America.

amplitude normalized to its value at high surface temperatures where $\theta \ll 1$. A plot of this quantity versus $1/T$ provides the binding energy of Na atoms on a glass surface, $E_b = 0.80$ eV. Further fitting of the dependence shown in Fig. 7.14 results also in a value of the polarizability of sodium adatoms, namely $\alpha_s = 6.6 \times 10^{-22}$ cm^3.

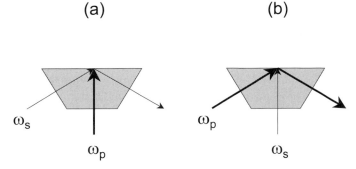

Fig. 7.15 (a) "Normal configuration" for studying scattering of sodium atoms at a glass surface. The pump laser excites a wave propagating normally to the prism surface which is resonant to the lower transition in the three-level cascade scheme. The probe laser excites an evanescent wave which scans across the upper transition. (b) "Inverse configuration". Here, the pump laser excites an evanescent wave, whereas the probe laser excites a wave propagating into the gas interior.

7.3.4
Excitation by Crossed Waves

An extremely small EW penetration depth into the gas permits one to excite gas atoms only within the Knudsen layer with a thickness of the order of the mean free path. Let us now investigate the possibilities that open up if one excites the boundary gas layer in addition to the EW wave, by an electromagnetic wave propagating normally to the surface. Then the optical response of the gas atoms is dictated by the dynamic polarizability, Eq. (7.22). If γ is much less than the Doppler width, $\alpha(\omega)$ has a sharp maximum for atoms with $v_z \approx \Delta/k_t$. Atoms moving to the surface and away from it have velocity components of different signs and thus one can distinguish between their contributions by tuning the laser frequency either below the atomic transition frequency ($\Delta < 0$) or above it ($\Delta > 0$). Therefore, simultaneous excitation of gas atoms by two crossed waves, one of which is evanescent, enables one to separate spectrally the contributions of the atoms immediately before the collision with the surface from those immediately after the surface scattering. The Doppler-broadened two-quantum fluorescence lineshapes are then determined by the velocity distribution functions of the atoms approaching the surface and departing from it and thus contain information about the gas–surface scattering dynamics.

The experimental setup for studying gas–surface scattering under those conditions implements a truncated prism to allow normal incidence of one of the beams onto the interface, as well as excitation of an EW with the other beam. In the experimental realization (Bordo et al. 2001) continuous wave lasers excited two adjacent transitions in Na atoms. In one excitation geometry (the "normal configuration"), the pump laser beam resonant to the

$3S_{1/2} \rightarrow 3P_{3/2}$ transition excited a wave normally incident onto the interface, whereas the probe (signal) laser beam excited an EW scanned across the $3P_{3/2} \rightarrow 5S_{1/2}$ transition (Fig. 7.15a). In the other excitation geometry ("inverse configuration") the character of the waves was exchanged (Fig. 7.15b). The atomic flux which was directed from the sodium source to the prism surface determined the velocity distribution function of the atoms arriving at the surface.

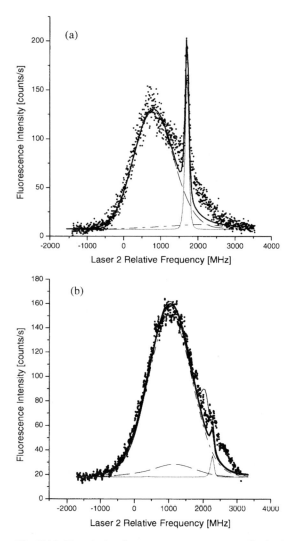

Fig. 7.16 Two-photon fluorescence spectra of sodium atoms observed in the "normal configuration". (a) $\Delta_p = -800$ MHz; (b) $\Delta_p = 800$ MHz. The contributions of different groups of atoms are shown separately: atoms directly from the sodium source (gray line), atoms scattered in front of the surface (dashed line), and atoms desorbed from the surface (dash-dotted line). The dark solid line is the complete fit curve. Reprinted with permission from Bordo et al. (2001). Copyright 2001, the American Physical Society.

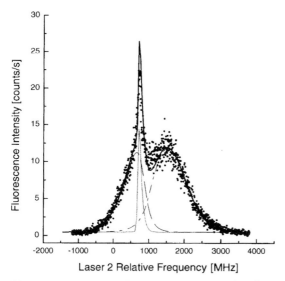

Fig. 7.17 Two-quantum fluorescence spectra of sodium atoms observed in the "inverse configuration". $\Delta_p = 200$ MHz. The notations are the same as in Fig. 7.16. Reprinted with permission from Bordo et al. (2001). Copyright 2001, the American Physical Society.

Typical two-photon fluorescence spectra obtained in the normal configuration are shown in Fig. 7.16. For negative detunings of the pump beam, Δ_p, a narrow peak is observed besides the Doppler-broadened line. This peak arises from the atomic flux emanating directly from the sodium source. The broader line originates from atoms arriving at the surface, too, but these atoms had undergone mutual collisions in front of the surface. For $\Delta_p > 0$ only atoms departing from the surface contribute to the spectrum. In the inverse configuration (see Fig. 7.17), the left and right wings of the spectrum arise from atoms arriving at the surface and departing from it, respectively.

Qualitatively, the difference between the spectra obtained in the inverse compared to the normal configuration can be explained as follows. The fluorescence intensity at a given probe beam detuning, Δ_s, is proportional to the number of atoms for which the probe beam is resonant, i.e., $\Delta_s - \mathbf{k}_s \cdot \mathbf{v} = 0$ with $\mathbf{k}_s$ the wave vector of the probe wave. In the normal configuration ($\mathbf{k}_s$ parallel to the surface) one scans the distribution function of atomic velocities along the surface by varying Δ_s. In contrast, in the inverse configuration ($\mathbf{k}_s$ perpendicular to the surface) the velocity distribution along the surface normal is scanned. In the latter case, the dip at the line center, $\Delta_s = 0$, results from Knudsen's cosine law: from Eq. (2.157) we can deduce that there are no atoms with $v_z = 0$ in the desorbing flux. This dip is washed out due to transit-time and power broadening. Thus, combining measurements in the normal and inverse configurations one can obtain comprehensive information on the

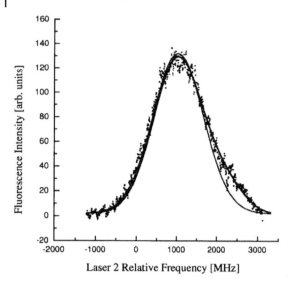

Fig. 7.18 The net contribution to the fluorescence spectrum from the sodium atoms scattered by the surface (dots). The calculated spectrum of the desorbed atoms and the best fit obtained for $\eta = 0.05$ are shown by the thin gray and black lines, respectively. $\Delta_p = 800$ MHz. The gray line does not fit the experimental data. η is the mean probability of direct scattering, which is low for the given experimental conditions. Reprinted with permission from Bordo and Rubahn (2003). Copyright 2003, the American Physical Society.

two-dimensional velocity distribution functions of gas atoms in the immediate vicinity of the surface.

A careful analysis of the fluorescence spectra obtained in the normal configuration permits extraction of additional important details of the gas–surface scattering process. We have seen that the contribution of the atoms departing from the surface can be separated for positive pump beam detunings at the lower atomic transition $3S_{1/2}(F = 2) \rightarrow 3P_{3/2}(F')$. However, for the nearby transition from the other hyperfine sublevel $3S_{1/2}(F = 1) \rightarrow 3P_{3/2}(F')$ Δ_p is negative. This leads to a small contribution to the spectrum from the atoms approaching the surface. Substracting this contribution gives the spectrum represented in Fig. 7.18. Here, the calculated fluorescence lineshape, assuming that all atoms moving from the surface have been desorbed from it (thin gray line), is also shown. A clear discrepancy between measurement and fit is observed, which can be attributed to the atoms which were scattered without being trapped at the surface, i.e., via a direct scattering channel. The velocity distribution of such atoms can be expressed in terms of the scattering kernel (see Section 2.4.2). Fitting the corresponding signal allows one to obtain the mean energy transfer to surface phonons as well as the mean probability of direct scattering, $\eta = 1 - S$ (Bordo and Rubahn 2003).

Further Reading

V.G. Bordo, H.-G. Rubahn, Evanescent Wave Spectroscopy in *Encyclopedia of Nanoscience and Nanotechnology*, Vol. 3, H.S. Nalwa (Ed.), American Scientific Publishers, California, 2004.

V.G. Bordo, H.-G. Rubahn, Multiphoton Spectroscopy Near Gas–Solid Interfaces in *Trends in Applied Spectroscopy*, Vol. 4, 2002.

M. Ducloy and M. Fichet, General theory of frequency modulated selective reflection. Influence of atom surface interactions. *J. Phys. II (France)* **1991**, *1*, 1429.

G. Nienhuis, F. Schuller, M. Ducloy, Nonlinear selective reflection from an atomic vapor at arbitrary incidence angle, *Phys. Rev. A* **1988**, *38*, 5197.

F. Schuller, O. Gorceix, M. Ducloy, Nonlinear selective reflection in cascade three-level atomic systems, *Phys. Rev. A* **1993**, *47*, 519.

M.F.H. Schuurmans, Spectral narrowing of selective reflection, *J. Phys. (Paris)* **1976**, *37*, 469.

Problems

Problem 7.1. Estimate the pressure of Cs vapor in a cell when the homogeneous broadening at the $6S_{1/2} \rightarrow 6P_{3/2}$ transition ($\lambda = 852$ nm) is comparable with the Doppler width at $T = 420$ K. The natural linewidth of the transition is $\gamma_n = 2\pi \times 5.3$ MHz, the collisional broadening is determined as $\gamma_c = 2\pi \times 1.2 \times 10^{-13} \times n$ MHz with n the number density of Cs atoms per cubic centimeter.

Problem 7.2. Cs vapor is kept in a cell at a temperature of $T = 420$ K and a pressure of $P = 0.05$ Torr. Estimate the angle of incidence of a light beam at the cell window–vapor interface, below which the spectral narrowing in selective reflection at the $6S_{1/2} \rightarrow 6P_{3/2}$ transition will occur. The index of refraction of the cell window is $n_w = 1.5$ and the other necessary parameters can be taken from Problem 7.1.

Problem 7.3. An evanescent wave is excited by a laser beam striking the prism–sodium vapor interface. The laser beam is resonant to the $3S_{1/2} \rightarrow 3P_{3/2}$ transition ($\lambda = 589$ nm, the natural linewidth is $\gamma_n = 2\pi \times 10$ MHz). Estimate the angle of incidence beyond which the fluorescence spectrum excited by the evanescent wave is sensitive to the change of the excited state population in vapor–surface scattering. Assume the index of refraction of the prism to be $n_p = 1.5$ and the vapor temperature to be $T = 300$ K.

Problem 7.4. Under the conditions of Problem 7.3, estimate the gas pressure below which evanescent wave spectroscopy is sensitive to a change in the vapor polarization in vapor–surface scattering. The collisional broadening of the transition is determined as $\gamma_c = 2\pi \times 0.70 \times 10^{-13} \times n$ MHz with n the number density of Na atoms in cm^{-3}.

Problem 7.5. The flux of Na atoms desorbing from the surface ($T = 400$ K) obeys the Knudsen law. One observes the excitation spectrum of fluorescence by scanning the laser frequency across the $3P_{3/2} \rightarrow 5S_{1/2}$ transition ($\lambda = 616$ nm) with the laser beam (a) parallel to the surface; (b) perpendicular to the surface, while the $3P_{3/2}$ state is populated by another laser. Calculate the frequency shift between the fluorescence line maxima in the two cases, provided that Doppler broadening dominates.

Note: Take into account that in this case the fluorescence lineshape resembles the shape of the distribution function, $f(v)$, over the atomic velocity projection, v, onto the wave vector of light, $\mathbf{k}$, i.e., it is determined by $f[(\omega - \omega_0)/k]$, where ω_0 is the atomic transition frequency. Then use the solution of Problem 2.9.

8
Optical Microscopy

8.1
Optical Resolution and Simple Light Microscopes

The use of light microscopes in their simplest form dates back to the ancient greeks. Figure 8.1 shows the setup of a light microscope of the kind that has been used since the 16th century after being introduced by Zacharias Janssen in 1590. The object is imaged with the help of an objective, and the image is magnified by an ocular. Thus the object image on the human retina appears larger compared with the image viewed without a microscope.

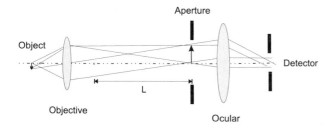

Fig. 8.1 A classical light microscope.

The magnification of the microscope is given by

$$V = V_{\text{objective}} \times V_{\text{ocular}} \tag{8.1}$$

where $V_{\text{objective}}$ is the magnification of the objective and V_{ocular} is the angular magnification of the ocular. Since, in most cases, a human eye looks through the microscope ocular, the ratio between the near-point of the human eye and the ocular focus length is used for V_{ocular}. The near-point is the shortest distance between the object and the human eye at which the eye is still able to focus. This distance strongly dependends on the age of the human individual. Hence one uses for convention a distance of 254 mm (10 inches).

Optics and Spectroscopy at Surfaces and Interfaces. Vladimir G. Bordo and Horst-Günter Rubahn
Copyright © 2005 WILEY-VCH Verlag GmbH & Co. KGaA, Weinheim
ISBN: 3-527-40560-7

Using this, we have for the magnification

$$V = -\frac{L}{f_{\text{objective}}} \cdot \frac{254mm}{f_{\text{ocular}}} \tag{8.2}$$

where f_{ocular} is the focal length of the ocular and $f_{\text{objective}}$ is the focal length of the objective. L is the tubular length, i.e., the distance between the focal point of the objective and the first focus of the ocular (usually 160 mm). A 10 times objective is thus an objective of focal length 16 mm and a 10 times-ocular has a focal length of 25.4 mm.

The smallest resolvable structure or the smallest distance between two separately imaged points is dictated by the wave nature of light. In the course of imaging a diffraction pattern is generated as shown in Fig. 8.2 for a point source. The corresponding intensity distribution is represented by an Airy function.

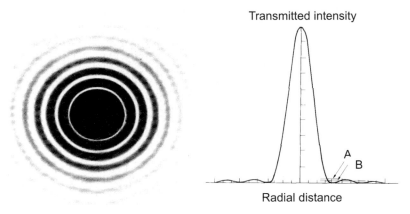

Transmitted intensity

Radial distance

Fig. 8.2 Optical resolution limit: diffraction pattern (Airy disc) and transmission function of a single point. Black colored regions denote maximum intensity. The abscissa is given in units of $\frac{2\pi}{\lambda} a \sin \alpha$ with λ the wavelength of the light, a the radius of the aperture and α half the viewing angle, as defined in Fig. 8.4. A and B denote the intensities of the first ($I_1 = 0.0175$ for $I(0) = 1.0$) and second maximum ($I_2 = 0.0042$). The radial distance of the first maximum from the zero point is $\frac{2\pi}{\lambda} a \sin \alpha = 5.14$ and that of the second, 8.42.

The diameter of the Airy disc for an imaging lens with focus length f and diameter D is

$$d_{\text{Airy}} \approx 2.44 \frac{\lambda f}{D} \tag{8.3}$$

Two points with diameter Δx can be separated, if the intensity maximum of one point falls into the intensity minimum of the second point ("Rayleigh–Abbe-criterion" (Abbe 1873; Rayleigh 1879) Fig. 8.3). This is the case for

$$\Delta x \approx 1.22 \times f \times \frac{\lambda}{D} \tag{8.4}$$

assuming that $\sin \alpha \approx \alpha$.

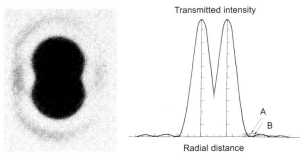

Fig. 8.3 Optical resolution limit: diffraction pattern (Airy discs) and transmission function of two neighboring points.

The minimum angular distance between two object points $\alpha_{min} = 1.22\lambda/D$ is called the resolving power. The resolving power increases with increasing diameter of the lens and with decreasing imaging wavelength. In the electron microscope, for example, wavelengths of 10^{-5} of the wavelength of visible light are used. The resolution of electron diffractive methods as compared with visible light diffraction is correspondingly higher.

If one uses a microscope objective instead of a simple lens, it is more appropriate to define the resolution limit as

$$d_{min} = k_1 \frac{\lambda}{NA} \tag{8.5}$$

In this equation, k_1 describes a coherence factor, which depends on exposure wavelength and varies between 0.55 and 0.8. In the case of an Airy-function (i.e., a point light source) one finds $k_1 = 0.61$. The factor NA is called the numerical aperture,

$$NA = n \cdot \sin \alpha \tag{8.6}$$

with n the index of refraction and α the aperture angle (Fig.8.4; 2α is the full angle of the focused beams).

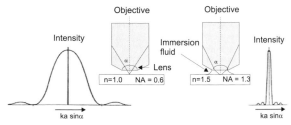

Fig. 8.4 Definition of the angle α, which is important for the numerical aperture NA, and two examples for the variation of the Airy diffraction image as a function of NA. The effective angle, under which the object is seen, amounts to 37° (left-hand side) and 58° (right-hand side).

The magnification $V' = \frac{0.12 \cdot NA}{\lambda[\text{mm}]}$ denotes the limit of useful magnification in a conventional microscope (tubus length 160 mm) since the diffraction and visual limit meet each other.

The numerical aperture (more exactly the square of the numerical aperture) describes the light collection efficiency of an objective and thus determines resolution and sensitivity. The values of numerical apertures above unity can be reached only if one takes advantage of total internal reflection. For that purpose, the objective and object holder are connected by a medium with high index of refraction (e.g., an immersion oil or water). In this way light is refracted via the immersion medium towards the normal of the objective (Fig. 8.4). Hence a larger number of Airy diffraction orders reach the imaging system and thus the contrast increases.

For practical applications the theoretical resolution limit (Eq. (8.5)) is in most cases not reached: imaging errors of the optical elements such as astigmatism, chromatic aberration or distortions determine the resolution limit. An additional limit is given by the limited number of pixels of the CCD cameras which are used in many cases instead of photo plates.

8.2
Dark-field, Fluorescence and Confocal Microscopy

An increase in the numerical aperture of the objective will increase the resolution that can be obtained with a given microscope. Illumination of the object to be investigated with a dark-field or evanescent wave condensor, results in a similar effect by increasing the contrast. Opposite to the conventional bright-field condensor (Fig. 8.5b) the dark-field condenser avoids illumination with the central light spot (Nachet 1847) (Fig. 8.5a). Hence illumination occurs mainly with light rays that propagate approximately parallel to the surface. The objective no longer collects the transmitted light rays emanating from the light source, but only the rays that have been reflected by the object. In this way the object appears bright in front of a dark background. The basic idea for this technique stems from 1903 (H. Siedentopf, R. Zsigmondy).

The evanescent wave condensor (Temple 1981) maximizes this principle by ensuring that the illumination occurs *solely* via surface waves, i.e., within the evanescent part of the electromagnetic field. Dark-field microscopy is also called ultramicroscopy since it allows one to investigate structures with characteristic dimensions smaller than the wavelength of the imaging light (visibility limit roughly $\lambda/100$).

As an example of contrast enhancement via dark-field illumination, bright- and dark-field images of the same object (a bent structure made of low contrast organic material) are shown in Figs 8.6a and b, respectively. As com-

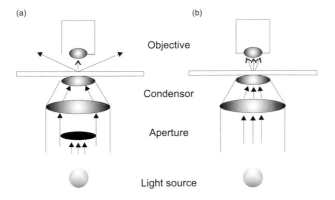

Fig. 8.5 (a) Dark-field condensor. (b) Bright-field condensor.

pared with bright-field illumination, dark-field illumination results in a clear contrast enhancement. The main reason for the strong contrast improvement is that the structures (needles in these examples) that are shown in Fig. 8.6 not only have a small diameter, but are also extremely flat (a thickness of less than 100 nm). They absorb only weakly in transmission. From the dark-field image an apparent diameter of a thick needle of about 1.3 μm is deduced. An atomic force microscopy image (Fig. 8.7b), however, reveals a true diameter of the needle of 500 nm.

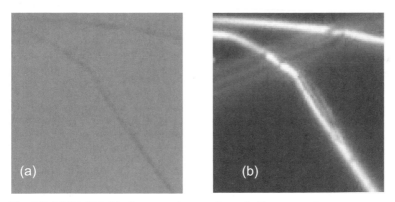

Fig. 8.6 (a) Bright-field microscopy image of needle-like aggregates made from organic molecules, adsorbed on mica. (b) Dark-field microscopy of the same area. The image size is 20×20 μm².

8.2.1
Fluorescence and Phase Contrast Microscopy

A further increase in resolution via an enlargement of the contrast ratio becomes possible if the object emits light, e.g., fluoresces. Figure 8.7a demon-

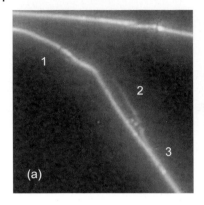 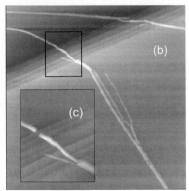

Fig. 8.7 (a) Fluorescence microscopy and (b) atomic force microscopy of the organic needle structures of Fig. 8.6. Image size $20 \times 20 \ \mu m^2$. Typical dimensions are: a break in the needle with width 300 nm (1), diameter of a thin needle 300 nm (2) and of a thick needle 500 nm (3). (c) Detail of break 1 (image size $5 \times 4 \ \mu m^2$).

strates this again with the help of organic needles (Fig. 8.6), which emit blue light after ultraviolet excitation. The substrate on which the objects have been grown is not emitting light, thus resulting in maximum contrast. The diameter of the thick needle (labeled 3) as determined from this image is about 800 nm. As detailed above, the real diameter, determined via atomic force microscopy (Fig. 8.7b) is 500 nm. Figure 8.7c shows a break in the needle during growth over a step edge on the mica substrate. The width of the break is about 300 nm. The thin needle structures possess diameters of the same order of magnitude, i.e., 300 nm.

As can be seen, it is possible via fluorescence microscopy to investigate structures with characteristic dimensions of a few hundred nanometers. In order to determine the real dimensions, however, complementary methods such as atomic force microscopy are necessary.

In the case of biological objects such as cytoplasm itself, the absorptivity in the visible spectral range is small, similar to that of the surrounding water. The indices of refraction, however, are slightly different ($n_{water} = 1.33$, $n_{cytoplasm} = 1.35$), resulting in a retardation of the incoming light wave. Following transmission through the object a phase difference exists with respect to the surrounding medium. This phase difference can be used for increasing the contrast as noted in the 1930s by Frits Zernicke (1953). A practical realization includes inserting a ring aperture between condenser and object holder, which avoids the direct light beam. A phase ring after the objective reduces the zeroth-order diffraction in order to enhance contrast. This leads to destructive interference of either the light waves that have been delayed (i.e., those that have transversed the cytoplasm) or those that have not been delayed. In the former case the object appears dark in front of a bright background, in the latter case the background is dark.

It has to be noted that, due to Abbe's diffraction limit, two objects cannot be individually observed which are separated by less than $\lambda/2$. However, if there is only a single object within the resolution limit the *position* of it can be determined with much higher precision. For this one has to scan spatially over the object, measuring the spatial intensity distribution. The precision of determining the position, δx, is then approximately given by

$$\delta \propto \frac{\Delta x}{SN\sqrt{n}} \tag{8.7}$$

where Δx denotes the width of the spatial intensity curve, SN the signal-to-noise ratio and n the number of measured points. This method has been used, for example, to determine the influence of the local environment on the properties of individual quantum dots with sizes much below the diffraction limit (Wu et al. 2000).

8.2.2
Confocal Microscopy

Scanning imaging methods in the optical range, especially near-field microscopy, usually result in a significantly higher resolution compared with that achievable using a conventional microscope. However, especially in biology and biophysics one is interested also in information about the interior of biological objects, e.g., living cells, as well as the surface or interface information. The introduction of confocal microscopy has enhanced the depth information significantly (Wilson 1990). Due to progress in dye-photochemistry and also quantum optics, fluorescence markers or more recently quantum dots attached to proteins, DNS and cell organells play a more and more important role. Such fluorescing or luminescing probes which possess high optical nonlinearities allow one to perform confocal microscopy via multiphoton processes. The introduction of multiphoton processes has the advantage that one avoids the necessity for UV-transmitting optical components and also that the resolution increases under certain circumstances (Schrader et al. 1998).[1]

Figure 8.8a shows the setup of a confocal microscope. The sample is illuminated from above via an aperture. Another aperture is placed in front of the photomultiplier (or another imaging system at the position of the ocular). This aperture serves to detect only rays which stem from the focal plane of the objective lens.

A variant of the confocal microscope is the laser scanning microscope (LSM, Fig. 8.9). Here, the effective focus acts as a three-dimensional probe, which is scanned in subsequent height steps two-dimensionally through a transpar-

---

1) To first order one would expect multiphoton excitation of fluorescence markers to decrease the resolution, since the excitation wavelength increases (Sheppard 1996).

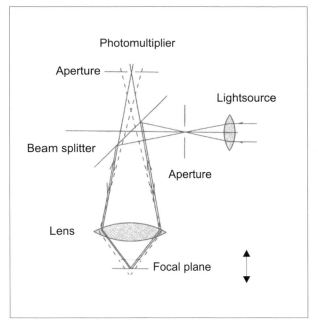

Fig. 8.8 Schematic setup of a confocal microscope. The ray propagation drawn with solid lines is for an object within the aperture diameter, the propagation with dashed lines for an object outside of the aperture.

ent object (Fig. 8.10). After scanning, the data are reconstructed into a three-dimensional image via sophisticated software.

As demonstrated in Fig. 8.11a with the help of various views of a blue light-emitting organic nanofiber, three-dimensional images of complex nanostructures can be obtained. Typical dimensions of the structure are a length of several micrometers, a width perpendicular to the surface of a few hundred nanometers and a width parallel to the surface of the order of micrometer. In fact, LSM is a good method of obtaining a view on the lower as well as the front-sides of nano-objects that have been grown on surfaces. The atomic force microscopy image of a similar nanostructure supported on a dielectric substrate in Fig. 8.11b shows that the details seen in Fig. 8.11a are not an artefact of the method. In order to avoid distortion of the nanostructure by the movement of the microscope objective, the nanofiber has been embedded in a highly viscous gel.

The limiting spatial resolution of confocal microscopy is demonstrated in Fig. 8.12a with the help of objects that have diameters far below the wavelength of the imaging light; here, fluorescing spheres of 110 nm diameter. The lateral resolution is of the order of a few hundred nanometers, while the vertical resolution of the order of a micrometer. This latter distortion of the spheres towards elongated objects results from the fact that the axial resolution is a factor of three to four smaller than the focal resolution.

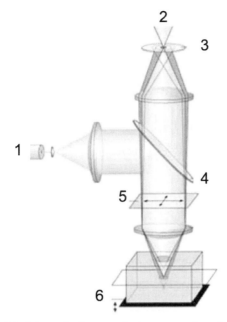

Fig. 8.9 Schematic diagram of a laser scanning microscope. 1, laser;
2, photomultiplier; 3, small aperture (pinhole) and beam splitter;
4, beam-splitter; 5, scanning mirrors; 6, focal plane, movable.

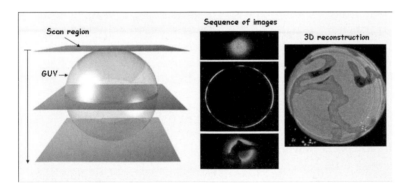

Fig. 8.10 Schematic of the scanning and reconstruction mechanism of a laser scanning microscope. The laser scans through a liposome (GUV, giant unilamellar vesicle, left), resulting in a fluorescent contour (middle), which can be reconstructed into a three-dimensional picture of the liposome (right). This liposome is formed by a mixture of fats which is similar to that of the human skin. (Bagatolli 2005).

    A way of improving the axial resolution is to illuminate the sample with coherent light from the back side. Coherent in this context means that a fixed phase relation exists between front- and back-side illumination. This can be achieved by illuminating from front- and back-side with light from the same laser source. In this "4π confocal microscopy" the axial higher order maxima

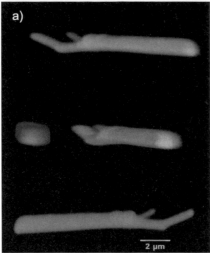
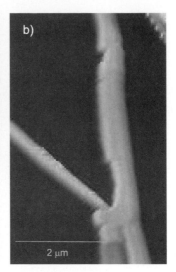

Fig. 8.11 (a) A series of confocal fluorescence microscopy images of a blue light-emitting nanofiber, embedded in a gel. The relative angles are 34°, 68°, 180° (right-hand side, from the top) and 90° (left-hand side). (b) Atomic force microscopy image of a similar nanofiber, but supported on a mica surface. (Brewer and Rubahn 2005).

in the scattering function are significantly reduced. Hence, the axial resolution becomes better (Fig. 8.12b). In the ideal case (complete illumination with solid angle 4π, which is of course strictly not possible with two lenses of finite sizes) the higher order maxima would be fully eliminated. However, with the help of subsequent image reconstruction this ideal case can be rather well simulated (Fig. 8.12c) (Schrader et al. 1998). Axial and lateral resolutions are then of the order of, or less than, 100 nm.

Another rather promising approach for biological applications in oreder to increase the far-field resolution's base on the use of two short pulsed lasers (Klar and Hell 1999). A first laser pulse (e.g., in the yellow spectral range around 558 nm) is used to excite the sample , while a second, picosecond time delayed pulse quenches the induced fluorescence (e.g., in the red spectral range around 766 nm). Quenching occurs in the overlap range of both pulses. The quench (STED – stimulated emission depletion) pulse is phase modulated such that it forms a spatial doughnut profile[2] around the excitation pulse. In this way the fluorescence image is confined to the inner core of the initial excitation spot and objects can be separated which are less than $\lambda/11$ apart. If one combines this trick with "4π confocal microscopy" then lateral and axial resolutions of less than 40 nm become possible.

---

2) A doughnut profile has a central intensity minimum with a spherical intensity maximum. The temporal overlap of a TEM_{01} and a TEM_{10} transverse mode results in just such an intensity profile.

(a) (b) (c)

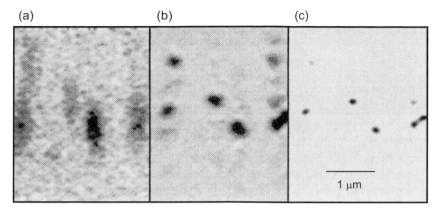

1 µm

Fig. 8.12 Images of a distribution of fluorescing spheres with 110 nm diameter, obtained via: (a) confocal microscopy; (b) 4π confocal microscopy; and (c) 4π confocal microscopy with image restoration. The image is shown in the (x,z)-direction, i.e., axial with the imaging focus. Reprinted with permission from Schrader et al. (1998). Copyright 1998, American Institute of Physics.

8.3
Total Internal Reflection Microscopy (TIRM)

Total Internal Reflection Microscopy serves, e.g., for obtaining the height distribution function between an – in most cases spherical – object and a surface from quantitative, time-resolved measurements of the intensity of light scattered within the evanescent wave. From the height distribution function one might find the interaction energy between object and surface.

Figure 8.13 shows the experimental setup: two microscope slides, separated by a spacer (e.g., a rubber ring) are optically contacted to a truncate (Dove) prism using an immersion liquid. In between the slides a liquid is filled which contains the spheres of interest. The whole setup is set below a conventional microscope objective and adjustable in (x,y,z)-coordinates. A laser beam enters the prism so as to obtain an angle larger than the angle of total internal reflection at the slide/liquid interface. Illumination both under normal incidence through the Dove prism and through the microscope objective is possible.

The instantaneous separation distance is determined by measuring the intensity of evanescent light scattered by the object. The intensity of the light flickers with time as the distance between the object and the lower microscope slide changes due to Brownian motion. Because of the nonuniform illumination of the liquid by the evanescent wave, the amount of light scattered by the object is a measure of its proximity to the interface. Let us measure the intensity change while the distance to the slide is varied. Then the intensity as

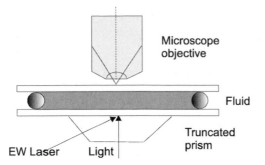

Fig. 8.13 Set up for total internal reflection microscopy.

a function of distance Δx can be described via

$$I(\Delta x) = I_0 \exp(-\beta \Delta x) \tag{8.8}$$

where β is given by

$$\beta = \frac{4\pi}{\lambda} \sqrt{(n_1 \sin \Theta_i)^2 - n_2^2} \tag{8.9}$$

i.e., it is the inverse penetration depth of the evanescent wave (Prieve and Walz 1993). As usual, n_1 and n_2 are the refractive indices of the microscope slide and fluid, Θ_i is the internal angle of incidence and λ is the wavelength of the evanescent light.

For this to be a quantitative method, one needs a value for I_0, the evanescent wave intensity at distance $\Delta x = 0$. This value can be obtained by forcing the object to contact the lower microscope slightly, via gravitational forces, changing the ionic strength of the solution, etc. Another important prerequisite of the method is that the height variations that are mapped via intensity variations should occur around a constant average height Δx within a measurement time of a few hundred seconds.

8.4
Brewster Angle Microscopy (BAM)

As shown in previous sections, p- and s-polarized light is diffracted at an interface between a surrounding medium and a substrate, differently as a function of angle of incidence. At an ideal interface, where the index of refraction jumps instantaneously from that of the medium to that of the substrate, p-polarized light is not reflected if the light hits the interface at the Brewster angle. At the Brewster angle, the reflected light beam and the light beam propagating into the substrate are at right angles to each other.

At a real interface the reflection of p-polarized light is minimized, but the reflection is not completely extinguished. Reasons for the remaining reflectance are the interface roughness, the finite thickness of the interface (i.e., the fact that $n = n(z)$, where z is the coordinate perpendicular to the surface) and possible anisotropies of adsorbed monolayers at the interface. This is used as the basis of Brewster angle microscopy (Hoenig and Moebius 1991; Hénon and Meunier 1991; de Mul and Mann 1998). A densely packed monolayer of amphiphilic molecules varies the index of refraction in the interface, $n(z)$, along a distance of $l = 2$ nm. Both, $n(z)$ and l depend on the different phases of the monolayers (e.g., solid vs. liquid) which leads to different depths of the Brewster minimum at these different phases and thus to an optical contrast. If one, for example, scans an argon ion or a helium–neon laser at the Brewster angle along the surface, a bright-dark image results, which reflects the different monolayer phases.

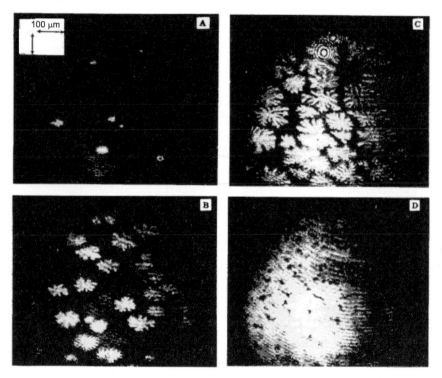

Fig. 8.14 Brewster angle microscopy images of DMPE on water as one compresses the film with a floating teflon barrier (A–D). Appearance and vanishing of solid domains is clearly observed. Reprinted with permission from Hoenig and Moebius (1991). Copyright 1991, American Chemical Society.

This is demonstrated in Fig. 8.14 for a monolayer of dimyristoylphosphatidylethanolamine (DMPE). The DMPE layer is floating on a water surface in a Langmuir–Blodgett trough. As the layer is compressed by a floating teflon barrier, solid domains appear (bright areas). This phase transition continues until no dark areas are any longer visible. Since a helium–neon laser has been used for imaging, no heating in the floating layer occurs and thus the phase transition is not affected by the imaging method.

In summary, this method is a two-dimensional variant of ellipsometry, which allows one to determine film thicknesses and optical properties of adsorbed monolayers. If the latter are known, one is able to deduce detailed morphological properties of the adsorbate films, such as molecular orientations or phase transitions. The lateral resolution is of the order of one micrometer, whereas thickness variations of the order of 0.1 nm can be deduced. The method is very different from phase-contrast microscopy since the phase domains themselves are imaged and not just the border lines (as in phase-contrast microscopy).

8.5
Phase Measurement Interference Microscopy (PMIM)

A dramatically increased resolution vertical to the surface is achieved by inserting a Mirau interferometer between microscope objective and sample (Fig. 8.15). Here, a tungsten halogen lamp is typically used as the illuminating light source. The light that is reflected from the surface to be investigated interferes with light reflected from a reference plate and the resulting interference pattern is imaged onto a CCD (charge coupled device) array. In fact, the Mirau interferometer consists of two small plates between the objective and the sample. One plate contains a small reflective spot that acts as the reference surface, and the other plate is coated on one side to act as a beam splitter. Next the interferometer is moved vertically with respect to the surface while the distance from the imaging lens to the reference surface remains constant. Hence, in one arm of the interferometer a phase shift is introduced while the generated interference pattern is measured by the CCD detector. This allows one to perform vertical scanning coherence peak sensing interferometry, i.e., to obtain a two-dimensional surface profile (Kino and Chim 1990; Wyant 1995).

Figures 8.16 and 8.17 show that the vertical resolution of a PMIM is comparable with that obtainable by the use of an atomic force microscope, namely better than 1 nm. As a test sample for these two figures a two-dimensional grating structure has been generated in an ultrathin film of poly-dimethylsiloxane (PDMS), adsorbed on glass. Both the substrate and the roughly one micrometer thick polymer film are nearly transparent, which makes optical imaging difficult. In addition, the PDMS film can be easily destoyed by mechanical forces so that a surface profilometry with a mechanical stylus becomes impossible. Tapping mode atomic force microscopy is a rather gentle way to obtain surface topological information as shown in Fig. 8.16a. The grating has a periodicity of roughly 4 micrometers with height variation of the order of 10 nm.

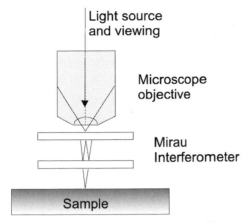

Fig. 8.15 Schematic drawing of an optical interference microscope. The Mirau interferometer can be moved upwards and downwards by a translational stage.

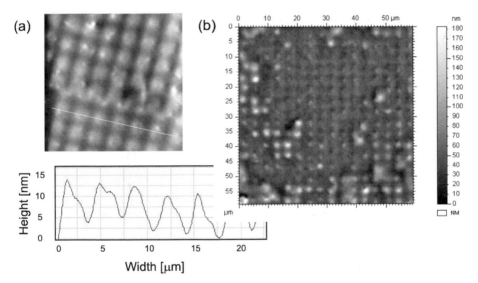

Fig. 8.16 A two-dimensional grating structure in a PDMS film as imaged by an AFM ((a), $25 \times 25 \ \mu m^2$) and an interference microscope (b). Copyright for the interference microscope image: Schaefer Technologie GmbH, 2004. Reprinted with permission. The line scan is a cross-section through the AFM image (white line), denoting the shallowness of the structure.

The PMIM measurement (Figs 8.16b and 8.17) provides the same local information but in addition allows one to easily obtain an overview over large surface areas of mm². There is no problem with adhesion of surface particles on the scanning tip, and height variation in the micrometer range can also be imaged. Of course, since it is an optical far-field method, the lateral resolution cannot be better than approximately 400 nm. Additional restrictions of the method are given on highly reflecting surfaces with low reflectivity contrast of neighboring areas.

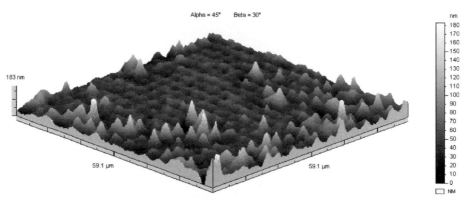

Fig. 8.17 Three-dimensional representation of the interference micro-scopical measurement of the surface structure from Fig. 8.16. Reprinted with permission from Schaefer Technologie GmbH, 2004.

8.6
Second Harmonic Microscopy

As shown in Section 6.2, optical second harmonic generation is a nonintrusive, selective and straightforward method to obtain surface- or interface-specific information. In the absence of noncentrosymmetric scatter sources, and given sensitive detectors, it can provide a very high signal-to-noise ratio and thus it is a rather obvious idea to obtain also two-dimensional SH or even sum frequency (SF) images (Floersheimer et al. 1997; Floersheimer 1999).

A typical experimental setup (Fig. 8.18) would consist of a scanning opti-cal microscope in reflection geometry on the base of a commercial microscope and a computer-controlled two-dimensional translation stage. The radiation source is typically a mode-locked pulsed Ti-Sapphire laser with a pulse du-ration of about 100 fs in order to avoid thermal damage of the sample and to increase the SH efficiency. The linearly polarized light beam from the laser would be used as a source of sample illumination at the fundamental har-monic frequency. An optical insulator is inserted at the laser output in order to avoid back-reflection into the laser cavity, and the half-wave plate along with the polarizer are used to control the polarization and power of the radi-ation incident on the sample. This setup then allows one to obtain polarized SH microscope images, thus allowing local determination of orientations of transition dipole moments on the surface; see below.

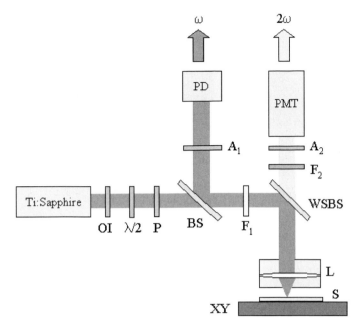

Fig. 8.18 Setup for SH microscopy. The abbreviations mean: PD: photodiode; PMT: photomultiplier; A_1: analyzer; F_1: filter; OI: optical isolator; P: polarizer; $\lambda/2$: half-wave plate; L: lens; S: sample; BS: beam-splitter; WSBS: wavelength selective beam splitter. Reprinted from Beermann et al. (2004) with permission. Copyright 2004, Elsevier Science.

After passing a wavelength-selective beam splitter, the laser beam would be focused at normal incidence on the sample surface with a long working distance objective of high numerical aperture. At the sample surface the full width at half maximum of the spot of the fundamental light is then about 1 micrometer, which gives a typical intensity at the surface of about 10^6 W/cm². The sample could be placed on the computer-controlled XY-piezoelectric stage, which is capable of moving in steps down to 50 nm with an accuracy of 4 nm over a scanning area of typically 25×25 mm². The sample reflects the fundamental and the second harmonic radiation, generated in the direction of reflection, back through the objective. The SH radiation is transmitted through a beam splitter and polarization analyzer and detected by a photomultiplier connected to a photon counter. The fundamental radiation is reflected out of the microscope and detected by a photodiode. During a normal scan, an area of 50×50 μm² would be scanned in 0.5 μm-steps with a speed of 30 μm/s, and the SH photons at each point are counted over a period of 20 ms, resulting in an image acquisition time of ten minutes.

Using this setup, the images in Fig. 8.19 have been obtained. The images (a) and (b) show the same surface area, which is covered by light-emitting nanofibers, but with differently polarized excitation light. Obviously, the op-

SH ←——————→

FH ←——————→

FH ↕

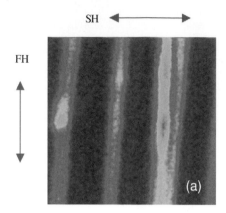

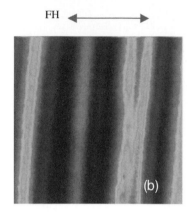

(a)

(b)

Fig. 8.19 SH microscopy (10×10 μm^2) of nanofibers for two polarization combinations. The excitation is with a femtosecond laser in the red spectral range, the detection is in the blue. a) Excitation light (first harmonic, FH) polarized along the needle axis (I_{sp}), b) ex- citation light polarized perpendicularly to the needle axis (I_{pp}). Detection (SH) polarized perpendicularly to the needle axis. Reprinted from Beermann et al. (2004) with permission. Copyright 2004, Elsevier Science.

tical response of the fibers depends locally on the polarization of the exciting light. From the ratio of the light intensities in the polarization combinations I_{sp} and I_{pp} one obtains, via

$$\frac{I_{sp}}{I_{pp}} = \tan^4 \theta \tag{8.10}$$

the angle $\pm\theta$ of the axis of the light-emitting molecules with respect to the polarization vector. Since the polarization ratio can be obtained spatially resolved from the microscopy images, it becomes possible to deduce optically the orientation of the molecules along those nano-aggregates with a precision of less than a micrometer. This information is difficult to obtain in any other way.

It is instructive to compare the SH microscopy results with the near-field images of Fig. 9.6 which show the same kind of light-emitting nanofibers. The spatial resolution of SH microscopy is lower (≈ 700 nm) compared with scanning near-field optical microscopy (SNOM); the contrast ratio, however, is much better. Nevertheless, it recently also became possible to determine molecular orientations via SH-SNOM (Fig. 9.14).

Further Reading

T. Wilson, *Confocal Microscopy*, (Academic, London, 1990).

B. Herman, *Fluorescence microscopy*, (BIOS Scientific, 1998).

E. Hecht, *Optics*, (Addison-Wesley, Reading, 2002).

Problems

Problem 8.1. If one wants to take advantage of the full resolving power of a lens or if one wants to focus light down to the minimum possible spot size, is this possible with any light source? Does one require special conditions?

Problem 8.2. Assume a laser beam being focused with a simple lens. How does the minimum irradiated area depend on the wavelength of the laser?

Problem 8.3. Optical microscopy in the far field is resolution-limited by the wavelength of the imaging light. What property of the light can be used to obtain resolution better than the wavelength and in which direction relative to the surface is this applicable for a far-field microscope?

Problem 8.4. If one wants to use either a Brewster angle microscope or an ellipsometer to determine the thickness of an adsorbate as one changes the temperature of the surface, which of them will give rise to the largest uncertainty?

Problem 8.5. Derive Eq. (8.10) taking into account that, for p-polarized exciting radiation, the transition dipole moment varies as $\cos^2 \theta$.

9
Nano-optics and Local Spectroscopy

Nano-optics deals with the understanding and the mastering of the zone between the micro- and macroworld using optical methods. In doing this, new optical properties are found which are based on the dimensional confinement that is a characteristic of nanoscaled materials. Metallic "quantum dots" such as Au nanoclusters are a good example of this domain, where changes in the size of the objects result in drastic changes in the optical properties (Kreibig 1997). These quantum dots have been well-studied in order to understand the fundamentals of the optoelectronic response in the nano-domain. In the meantime, they have already been used, e.g., for enhancing the brightness and stability of fluorophores for biological imaging and are now commercially available. This illustrates the speed of transfer of basic research results to industrial products in this field.

Another example, based on dielectric materials, are photonic band gap (PBG) materials (Joannopoulos et al. 1995). Here, a periodic index modulation is manufactured in dielectric slabs (e.g., by laser- or electron-beam-drilling a matrix of submicron-sized holes) which in the following inhibits the propagation of light of certain colors over a large range of scattering angles. In analogy with terms from solid state physics this is called an "optical band structure". Again, the commercialization occurred in the very short timescale of a few years, and optical fibers implementing the PBG effect are now available for a wide range of applications. In contrast to the above-mentioned quantum dots, the PBG effect can be quantitatively understood using classical electrodynamics. And indeed the possibility of modeling the optical behaviour of submicron-scaled materials using classical methods is often encountered in the context of nano-optics.

The above examples from metallic and dielectric nanoscaled systems should not give the impression that optics in the sub-wavelength size regime is well understood. The influence of morphological changes on a nanometer-scale on the optical near-field and from that on the resulting far-field needs to be investigated as well as the corresponding influence on the spectroscopic properties

Optics and Spectroscopy at Surfaces and Interfaces. Vladimir G. Bordo and Horst-Günter Rubahn
Copyright © 2005 WILEY-VCH Verlag GmbH & Co. KGaA, Weinheim
ISBN: 3-527-40560-7

or the intrinsic dynamics of optical excitations in nano-aggregates. On the application side, a thorough understanding of this dependence is an important prerequisite for the controlled build-up of new nanoscaled, surface-based optoelectronic elements such as light-emitting devices or field-effect transistors.

In this chapter we discuss some recent developments in near-field imaging and spectroscopy.

9.1
Scanning Near-field Optical Microscopy (SNOM) and Photon Scanning Tunneling Microscopy (PSTM)

A widely distributed and very useful synthesis of atomic force microscopy (AFM) and conventional optical microscopy is scanning near-field optical microscopy (SNOM). The principle of operation, advantages and limitations of this new kind of microscopy are very much related to "conventional" scanning microscopies such as scanning tunneling microscopy (STM). In passing, we note that STM soon celebrates its 25th anniversary and AFM its 20th anniversary. Accordingly, both methods are very mature nowadays and a wealth of modification for specific purposes have been developed (Wiesendanger 1994; Sarid 1997; Wiesendanger 1998). The SNOM is one of these.

Scanning tunnel microscopes allow a vertical resolution in the sub-nanometer or even sub-Ångstrom regime over large areas on the surface (Chen 1993; Fuchs 1994). In the STM (Binnig et al. 1982) changes in the tunnel current j_T between a conducting tip and a conducting surface as a function of distance between tip and surface, form the basis of the imaging principle. Due to the exponential distance dependence, even a small change in distance between tip and surface results in a strong change of tunneling current. Hence the high sensitivity of the method. The tip is mounted onto piezoelectric elements, which allow reproducible movements in the sub-nanometer regime. The precision of movement is dominated by the noise of the amplifiers and is typically of the order of $0.4 \, \text{pm}/\sqrt{\text{Hz}}$ (Hicks and Atherton 1997).

The measurement of a tunneling current relies on the existence of free charge carriers in the investigated substrate. While there are numerous STM studies on metal and semiconductor surfaces (e.g., Besenbacher (1996)), it is rather difficult to investigate insulator surfaces with an STM. An exception to this are ultrathin isolating films (e.g., organic films on metals), which allow tunneling of the electrons from the metal to the tip. If the films follow their typical growth behaviour on metals, i.e., they begin to build nano-aggregates, however, the imaging with an STM becomes more and more difficult. Here, either optical methods or atomic force microscopy can be applied. The AFM (Binnig et al. 1986) is based on the use of the van-der-Waals force between tip and surface, which leads to vertical and torsional deflections of the tip, which is mounted on a cantilever. This force decreases with the third power as a function of distance to the surface (Magonov and Whangbo 1996).

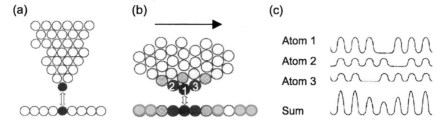

Fig. 9.1 A scanning tunneling microscope probes local atomic interactions (a), while the cantilever deflection of an atomic force microscope has a lower order distance dependence (b). A defect in the surface (e.g., a missing atom) vanishes easily in the periodic background (c): the local resolution is less than that of a STM. Reprinted with permission from Rubahn (2004). Copyright 2004, B.G. Teubner Verlag.

For a quantitative analysis of AFM images in the nanometer regime the diameter of the AFM tip has to be taken into account, which amounts to several (or even several tens of) nanometers. Since the deflection of the cantilever is induced by interactions of several atoms (not only a single atom as in the case of an STM) the possible resolution of an ATM is much lower compared with that of a STM (Fig. 9.1). This is even the case if the AFM has a super-sharp tip and very much limits the topographical resolution of a SNOM. Periodic structures can be imaged with atomic resolution, but not isolated details. In order to correct this effect, the image has to be deconvoluted with the effective tip shape (Todd and Eppell 2001). The shape of the actual tip has to be determined by imaging of a sample surface with very well known topology (e.g., a periodic array of nanometer-scaled particles) and following deconvolution. Even under optimum conditions this method is nontrivial and not unique. Simultaneous measurements using an alternative technique (e.g., fluorescence microscopy) are therefore very useful.

For this purpose, the combination of an inverted optical microscope with an atomic force microscope on top has proved very useful (Fig. 9.2), especially in the biological sciences, where AFMs are nowadays a very important experimental tool (Morris et al. 1999). Note that the most versatile solution is an AFM which allows optical imaging also from the top via, e.g., phase contrast microscopy. This setup gives access both to transparent objects (fluorescence imaging from below) as well as nontransparent materials and is therefore also very well suited for simultaneous morphological and optical imaging of submicron scaled circuits and other elements of new solid state electronics.

Essentially, the SNOM is based on the idea that the diffraction limit of the optical microscope can be broken if the imaging aperture or the detector are positioned in the *near field* of the reflecting or emitting object. This idea dates

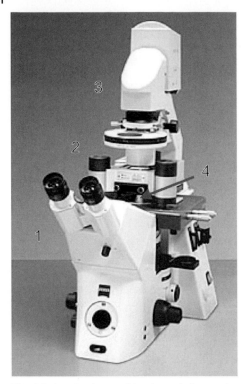

Fig. 9.2 Image of a combined atomic force microscope/inverted micro-scope setup (NanoWizard in the LifeScience version, with permission from JPK Instruments AG, Berlin). The numbers denote: 1, inverted microscope; 2, AFM; 3, top illumination and possibly view stage; 4, sample.

back, in fact, to the twenties of the last century (Synge 1928). It has been realized experimentally in the seventies by a reflection-SNM (scanning near-field microscope) for centimeter waves (Ash and Nicholls 1972) with a resolution of $\lambda/15$. Ten years later a variant in the visible spectral range ($\lambda = 488$ nm) of the scanning near-field optical microscope, was realized with a resolution of better than 25 nm or $\lambda/20$ (Pohl 1982; Pohl et al. 1984; Pohl and Courjon 1993). At the end of the eighties a new modification of the SNOM was constructed: the PSTM (photon scanning tunneling microscope, Figs. 9.3 and 9.7) (Reddick et al. 1989).

The most important ingredient of the SNOM, of course, is that the imaging aperture stays in the near field of the object and that it can be moved with a precision of nanometers along the surface. The development of scanning microscopes and especially of the AFM as described above made this technology available and paved the way to the SNOM (Paesler and Mojer 1996). Instead of an imaging aperture one usually implements a sharpened optical

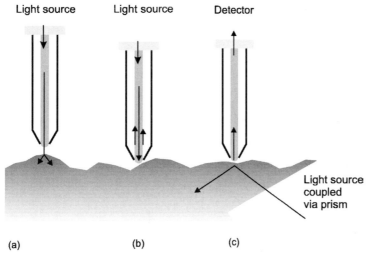

Fig. 9.3 Imaging via near-field microscopy: (a) SNOM in transmission; (b) SNOM in reflection; (c) PSTM.

fiber, which is coated with an aluminum cover into which a circular aperture of a few ten nanometers has been eroded. In this way the collected photons from the investigated object can be transferred directly to a photomultiplier. The optical fiber is usually connected with a piezo stack and, just as in the case of a STM, shear forces are measured and used as the feedback signal during scanning. The read-out of this signal allows one to obtain topographical information simultaneously with the optical information. The resolution limit is given by the aperture size, which cannot be infinitely small (Fig. 9.4).

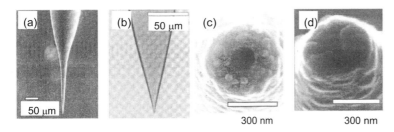

Fig. 9.4 Side- and front view of aluminum-covered optical glass fibers. The images (a) and (c) as well as (b) and (d) are related to each other. (b) is a light microscope image, (a), (c) and (d) are SEM images. The SNOM aperture is visible as a dark circle in (c). Reprinted with permission from Hecht et al. (2000). Copyright 2000, American Institute of Physics.

Figure 9.5 shows a metal-coated SNOM tip above an ensemble of light-emitting nanostructures, observed via a scanning electron microscope (SEM). There is an obvious size difference between detector and nanoscopic objects, which results in a low detection sensitivity and a huge uncertainty in the determination of the position of the tip. In Fig. 9.5 the light-emitting objects are positioned on an electrical insulating surface in order not to disturb their optical properties via the surface. Usually insulating surfaces are difficult to image with a SEM due to charging effects. To overcome this problem, a weak water vapor pressure has been added to the SEM chamber, which slightly decreases the resolution (ESEM, environmental scanning electron microscope).

(a) (b)

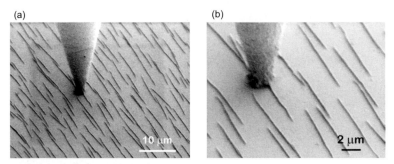

Fig. 9.5 (a) SEM image of a SNOM tip above an array of light-emitting nanofibers. Reprinted with permission from Sturm et al. (2003).
(b) Higher resolution image of the same tip.

Figure 9.6 shows simultaneously obtained topographical and optical images of nanoscopic light-emitting objects (nanofibers). The topography (Fig.9.6a) follows directly from the feedback signal that allows constant height-scanning of the tip over the surface. In order to obtain an optical signal, the nano-aggregates had to be excited with ultraviolet light. Depending on whether the SNOM scan occurs perpendicular to the long needle axis (9.6b) or parallel to it (9.6c), glowing of the excited objects is observed (9.6b) or waveguiding of light along some of the needles (9.6c)[1]. Waveguiding is seen as glowing of the needles along their whole length since the SNOM samples the photons in the near field of the needle – the SNOM acts as a defect on the needles, which leads to light scattering.

The resolution limit of the SNOM is theoretically improvable towards molecular resolution by avoiding damping through the metal coating of the fiber. This is called "apertureless" near-field optics (Zenhausern et al. 1994). A possibility for this is the PSTM (Fig. 9.7), i.e., the excitation of the sample surface via an evanescent wave and detection by dipping the fiber tip into the near field of the surface (Fig. 9.6 has been obtained this way). As an alternative, the near-field limited field might also be generated via far-field

---

1) The excitation has been done with unpolarized light. Only light that is polarized perpendicular to the long needle axes is used for excitation since the transition dipole moments of the molecules from which the needles are built are oriented perpendicular to the needle axes.

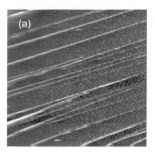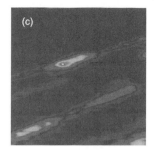

Fig. 9.6 SNOM images (40×40 µm^2) of nanofibers. Images (a) and (c) have been obtained simultaneously, (a) shows the topography of the aggregates, (c) the optical response after exciting the nanofibers. In (b) the SNOM is scanned perpendicular to the long needle axes, in (c) parallel to them. Waveguiding occurs only along the needle axes. Reprinted with permission from Volkov et al. (2004). Copyright 2004, The Royal Microscopical Society.

illumination of a strong scatterer, which is in direct optical contact with the fiber which transmits the light to the detector. Thus, within this method, fiber and scatterer are scanned in a small distance over the surface, which is illuminated in the far field.

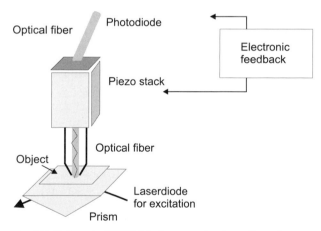

Fig. 9.7 Setup of a PSTM (photon scanning tunneling microscope). Localized electromagnetic fields in the near field of the sample surface are detected with the help of a noncoated dielectric tip. Reprinted with permission from Rubahn (2004). Copyright 2004, B.G. Teubner Verlag.

While the resolution is no longer restricted by the damping aperture, an obvious disadvantage of this method is that small signal intensities are measured in front of a high background intensity (the illuminated surface). Thus the method only works reasonably well if the scattering rate of the object that has been mounted to the SNOM tip is strongly enhanced, e.g., via plasmon excitation (Fischer and Pohl 1989).

Eventually the scattering object at the tip of the fiber could also be a single, fluorescing molecule or single excited center (Sekatskii and Letokhov 1996). In most cases one does not attach individual molecules at the optical fiber but doped host crystals. If it becomes possible to identify spectroscopically individual molecules in this host crystal, one should be able to use them as single-molecule light sources for optical near-field microscopy. Figure 9.8 is the setup for an experimental realisation of this idea (Michaelis et al. 2000).

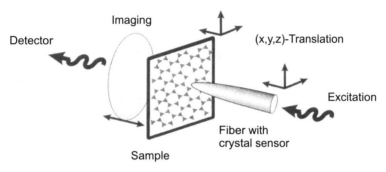

Fig. 9.8 Setup for optical microscopy with individual molecules. Reprinted with permission from Michaelis et al. (2000). Copyright 2000, Nature Publishing Group.

As a host crystal a p-terphenyl crystal, with a few micrometers in diameter, has been used, which has been doped with a low concentration (10^{-7}) of terrylene molecules. The tip of the optical fiber has been cooled to 1.4 K in order to avoid broadening of the fluorescence line via collective phenomena such as interaction with the phonons of the host crystal. The observed linewidth in fact is the natural linewidth (a few tens of MHz; Fig. 9.9). The spectral response of these individual molecules can now be used for a local analysis of the surface, albeit with relatively low resolution (180 nm) (Fig. 9.10). Further experimental tricks such as electrical field induced Stark shifts can be applied to determine at least the position of the investigated molecules to within a few 10 nm.

For the images shown in Fig. 9.10, the sample consisted of a microscope plate, which has been decorated with a hexagonal lattice of 25 nm high, triangular aluminum islands. The lattice period was 1.7 μm. In the force microscopy image (Fig. 9.10a) two differently oriented aluminum triangles are visible (*i* and *ii*). The aluminum triangles suppress the optical signal and appear as dark spots in the optical image (Fig. 9.10b). Differently oriented triangles can easily be identified.

The single molecule light source is based also on the idea of performing field ion microscopy with molecules that have been marked via laser light. This

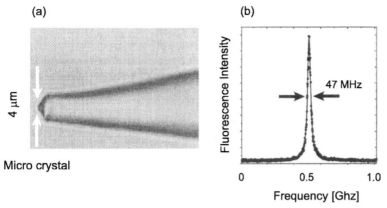

Fig. 9.9 (a) Light microscope image of an optical fiber to which a terrylene-doped p-terphenyl host crystal has been attached. (b) Excitation spectrum of the terrylene molecule at $T = 1.4$ K. Reprinted with permission from Michaelis et al. (2000). Copyright 2000, Nature Publishing Group.

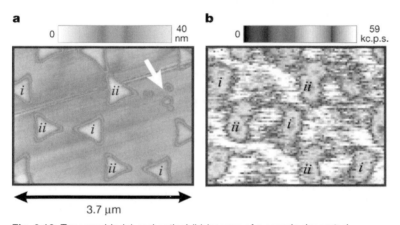

Fig. 9.10 Topographic (a) and optical (b) images of a sample decorated with a hexagonal aluminum lattice. The optical image has been obtained in the light of a single terrylene molecule. Reprinted with permission from Michaelis et al. (2000). Copyright 2000, Nature Publishing Group.

would allow one to localize spatially individual bonds on a surface (Letokhov 1975; Chekalin et al. 1984). If one mounts a donor dye molecule on the surface and acceptor molecules at the tip of a SNOM or AFM, one is able to obtain, not only optical information about the surface topography, but also about the resonant energy transfer processes (Shubeita et al. 1999) (FRET, fluorescence resonant energy transfer).

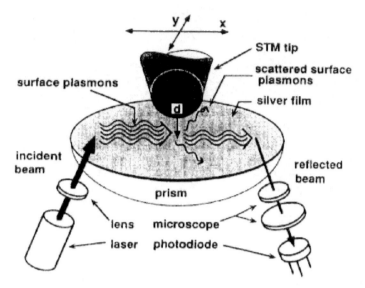

Fig. 9.11 Schematic drawing of a scanning plasmon near-field microscope. A STM tip modifies the propagation of surface plasmons on a thin silver surface. Scanning of the tip results in a two-dimensional image of the surface. Reprinted with permission from Specht et al. (1992). Copyright 1992, American Physical Society.

9.2
Scanning Plasmon Near-field Microscopy (SPNM)

Scanning near-field optical microscopy can be used to obtain detailed images of localized surface plasmons, e.g., around lithographically fabricated gold dots (Hecht et al. 1996; Kim et al. 2003). The resolution, however, is limited by the small detection sensitivity, which limits the minimum aperture size of the scanning optical fiber. Usually one obtains a resolution of the order of 100 nm.

One way of drastically increasing the lateral resolution down to a few nanometers is given by the combination of a STM tip (radius of curvature around 10 nm) and resonantly excited extended surface plasmons (Specht et al. 1992). For this technique, surface plasmons are excited in a thin metallic film on a prism (Fig. 9.11). They are monitored by the intensity change of the totally internally reflected exciting laser beam, which shows a pronounced minimum as a function of the angle of incidence at the angle where surface plasmons are excited (see also Fig. 3.8).

The depth of this minimum is modified as the STM tip penetrates the evanescent field above the surface, and this modification is measured as the tip is scanned two-dimensionally over the surface. The method is especially sensitive due to radiationless energy transfer from the tip to the film, which has a cubic distance dependence. In contrast to plain STM imaging, SPNM bears spectroscopic capabilities due to the light coupling and is a possibly more gentle method since the distance between tip and surface is of a the order of a few nanometers (non-contact). A severe disadvantage is certainly the need to excite surface plasmons which very much restricts the number of systems that can be investigated. In addition, inherent background noise limits the minimum measurable signal changes to a few percent and thus restricts the sensitivity of the method.

9.3
Near-field Optical Spectroscopy

9.3.1
Fluorescence Spectroscopy

Besides imaging it would be of great help, for a thorough understanding of light generation, propagation and transformation in and along nanoscaled structures, if one were able to perform spectroscopy in the near field. Of course, besides severe limitations due to the small signal-to-noise ratio given by the inefficient coupling from the near field into propagation modes of a SNOM fiber, the presence of a metallic aperture would affect such spectroscopic information severely. Thus, an apertureless approach is more appropriate.

One method is to scatter light from metallic nanostructures and employ localization and enhancement effects via plasmon coupling (plasmonics) as discussed briefly above. Or the metal tip, illuminated by far-field light, can provide a localized excitation source for the sample that is investigated spectroscopically. A variant of this is to laser-trap metallic particles of several tens of nanometers width and to use them as near-field probes[2] (Kawata et al. 1994). After trapping the particles with an infrared laser beam, it is moved close to the surface and a visible laser beam is scattered there. This latter beam excites surface plasmons, which results in a strong local field enhancement. This in turn is used as a point light source (a few ten nanometers in width) to excite fluorescence in the sample. The fluorescence is measured in the far field, but scanning the near-field excitation source with nanometric resolution provides optical resolution far below the diffraction limit. Scanning of the laser-illuminated tip itself, also provides near-field excitation of surface structures, the response of which is detected at other wavelengths in the far fied (Sanchez et al. 1999).

---

2) This is very much a realization of the basic idea of Synge, from 1928, for ultramicroscopy.

Figure 9.12 shows, as an example, topographic and near-field spectroscopic images of dye aggregates embedded in a polymeric surface. From the line scans a spatial resolution of about 30 nm can be deduced.

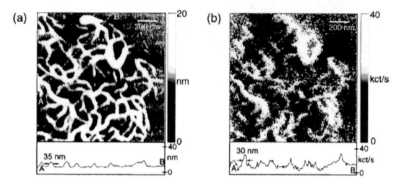

Fig. 9.12 Topographic (a) and near-field two-photon fluorescence (b) image of J-aggregates of a dye in a polymeric film on glass. In the lower part of the figure height scans through the images are shown. Reprinted with permission from Sanchez et al. (1999). Copyright 1999, American Physical Society.

For these prototypical experiments, two-photon excitation via femtosecond laser illumination has been used in order to improve the contrast ratio and avoid a strong background signal from scattered light. Depending on the magnitude of the shift between excitation and emission light frequencies, however, these experiments should also be performable with single-photon excitation. A further problem might be, in general, that the metal tip quenches the fluorescence, which obviously competes with the local field enhancement effect. In addition, systems with high fluorescence quantum yield have to be employed in order to obtain a satisfactory signal-to-noise ratio.

9.3.2
Raman Spectroscopy

A further development of the method described in the above section is to avoid fluorescence imaging and instead to use the Raman scattered light from the surface (Hartschuh et al. 2003). Raman scattering probes the vibrational spectrum of the sample, thus sensitively reflecting chemical composition, molecular structure and their changes due to the surface environment. The low Raman scattering cross-section (around 14 orders of magnitude lower than that for fluorescence emission) is rather straightforwardly overcome by an enhancement of the same order of magnitude in the case of light scattering by appropriate surface roughness (surface enhanced Raman scattering, SERS). Conversely, scattering of a sharp metal tip can also dramatically enhance the Raman scattering intensity. Maximum field enhancement is achieved for polarization of the exciting laser beam along the tip axis.

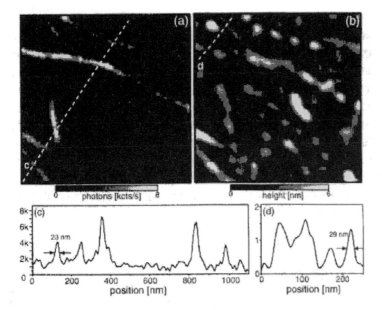

Fig. 9.13 Near-field Raman (a) and topographical (b) image of carbon nanotubes on glass. The Raman image is for a specific Raman line corresponding to a tangential stretching mode of the carbon nanotubes. In the lower part of the figure height scans through the images are shown. Height in nm for the topography. Reprinted with permission from Hartschuh et al. (2003). Copyright 2003, American Physical Society.

The achievable spatial resolution of this method of about 25 nm is demonstrated in Fig. 9.13 with the help of a near-field Raman image (Fig. 9.13a) and a topographical image (Fig. 9.13b) of single-wall carbon nanotubes on glass. It is interesting to note that humidity-related circular features in the topographical image are not seen in the Raman image since they do not provide a SERS signal. This demonstrates the chemical specificity of the method.

We note that surface-enhanced infrared absorption can also be used for near-field probing of chemical constitution with a local resolution of less than 100 nm (Knoll and Kellmann 1999). In fact, fortunately one finds in many different cases, surface enhancement effects caused by the scanning tip, which should allow one to use a variety of traditionally far-field spectroscopies in the near-field regime.

9.4
Near-field Nonlinear Optics

In addition to field enhancement via localized plasmon excitation, nonlinear optical methods, such as optical second harmonic generation (SHG), deliver significant contrast enhancement which can be used for near-field experiments. This has been demonstrated recently with the help of SHG generation from InAlGaAs semiconductor quantum dots on GaAs(001) both in the far field (SH-SFOM, SH-scanning far field optical microscopy, (Erland et al. 2000)) and in the near field (SH-SNOM, (Vohnsen et al. 2001)). SH-SNOM works without metal coating of the optical fiber but the excitation of the sample occurs in the near field of the fiber. Hence this SNOM modification is best described by Fig. 9.3a.

Let us assume that we obtain a two-photon image induced in the near field (similar to Fig. 9.12 from the nanoscaled organic fibers that are shown in Fig. 8.19 as far-field images. If we measure two sets of data under different polarization combinations of exciting and detecting light we might obtain near-field information about the orientation of molecular transition dipole moments.

Figure 9.14 shows two-dimensional plots of such orientations, obtained from a direct comparison of differently polarized TP-SNOM (two-photon SNOM) images of luminescent organic nanofibers. The resolution is better than 400 nm, at least a factor of two better than the comparable far-field results that can be obtained from the ratio of the two images in Fig. 8.19. From this image it is observed that the molecules which build the nanofibers are rather homogeneously oriented along the nanoscaled objects.

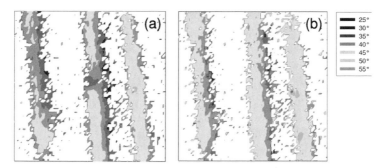

Fig. 9.14 Two-dimensional images of molecular orientations along light-emitting nanofibers, obtained from polarized TP-SNOM measurements and applying Eq. (8.10). The two images show the same aggregates with measured data obtained for sp/pp orientations and ss/ps orientations, respectively. The angles on the right-hand side correspond to the orientation of the individual molecules with respect to the short axis of the needle-like aggregates. (Beermann et al. 2005)

Further Reading

V.A. Markel and T.F. George, Eds. *Optics of Nanostructured Materials*, (John Wiley & Sons, New York, 2001).

S.Kawata, M. Ohtsu and M. Irie, Eds. *Nano-Optics*, (Springer, Berlin, 2002).

D. Courjon, *Near-Field Microscopy and Near-Field Optics*, (Imperial College Press, London, 2003).

P. N. Prasad, *Nanophotonics*, (Wiley, Hoboken, 2004).

Problems

Problem 9.1. If one is to characterize the near and far field by a relationship between the characteristic size of an object and the wavelength of light, what kind of relationship would be useful? What kind of diffraction theory has to be applied in the near and far field, respectively?

Problem 9.2. In a SNOM with a metal-coated fiber, what limits the minimum size of the aperture and thus the near-field resolution?

Problem 9.3. What determines the topographical resolution of a SNOM and why is it usually so bad?

Problem 9.4. Raman scattering is strongly enhanced at the tip of a scanning STM. What is the basic physical effect behind this observation?

10
Solutions to Problems

Problem 2.1. The arrangements of atoms in the (111) surface of bcc and fcc crystals as well as the directions of the $[\bar{1}10]$ and $[\bar{1}\bar{1}2]$ axes are shown in Figs 10.1 (a) and (b), respectively. The surface unit cells are indicated by dashed lines. In both cases the surface lattices have hexagonal symmetry.

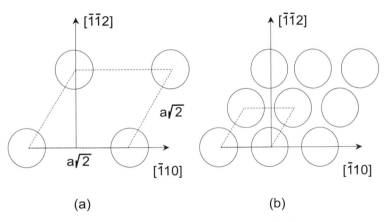

Fig. 10.1 Arrangement of atoms in the (111) surface of (a) a bcc crystal; (b) a fcc crystal. a is the lattice constant.

Problem 2.2. The electronic surface states are shown in Fig. 2.26 by dashed lines. The parabolic shape of the upper surface band implies that it is well described by the nearly-free electron approximation and hence this state is classified as a Shockley state. The lower surface state has the character of a surface resonance in the region where it intersects the electronic bulk band. An electronic transition between the surface states is possible if the upper state is unoccupied, i.e., it is located above the Fermi level. The minimum energy of such a transition is about 1.8 eV.

Optics and Spectroscopy at Surfaces and Interfaces. Vladimir G. Bordo and Horst-Günter Rubahn
Copyright © 2005 WILEY-VCH Verlag GmbH & Co. KGaA, Weinheim
ISBN: 3-527-40560-7

Problem 2.3. The dispersion curves of surface phonons are shown in Fig. 2.27 by thick solid and dashed lines. Those which are indicated by dashed lines have the character of surface resonances. The branch of acoustical surface phonons (S_1) originates at the $\bar{\Gamma}$-point ($\mathbf{k}_\| = 0$) where its energy is equal to zero. The dispersion curves S_2 and S_3 correspond to optical surface phonons. One cannot classify the curve S_4 based on this figure.

Problem 2.4. According to Eq. (2.91) the power spectrum is determined by the Fourier transform of the surface profile

$$\hat{\zeta}(k_x, k_y) = h \lim_{X,Y\to\infty} \int_{-X}^{X} \int_{-Y}^{Y} \cos\left(\frac{2\pi}{a}x\right) \cos\left(\frac{2\pi}{b}y\right) e^{-ik_x x} e^{-ik_y y} dx\, dy \quad (10.1)$$

A direct evaluation together with the use of Eq. (2.32) leads to four terms of the form

$$\lim_{X,Y\to\infty} \frac{\sin\{[k_x \pm (2\pi/a)]X\}}{k_x \pm (2\pi/a)} \cdot \frac{\sin\{[k_y \pm (2\pi/b)]Y\}}{k_y \pm (2\pi/b)} \quad (10.2)$$

$$= \pi^2 \delta\left(k_x \pm \frac{2\pi}{a}\right) \delta\left(k_y \pm \frac{2\pi}{b}\right) \quad (10.3)$$

Now the power spectrum can be found as

$$P(k_x, k_y) = \lim_{X,Y\to\infty} \frac{1}{4XY} \left[\hat{\zeta}(k_x, k_y)\right]^2 \quad (10.4)$$

Then substituting the representation of $\hat{\zeta}(k_x, k_y)$ in the form of Eq. (10.2) instead of one power, and its representation in the form of Eq. (10.3) instead of the second power, one finally obtains

$$P(k_x, k_y) = \frac{\pi^2 h^2}{4}$$
$$\times \left[\delta\left(k_x - \frac{2\pi}{a}\right) \delta\left(k_y - \frac{2\pi}{b}\right) + \delta\left(k_x - \frac{2\pi}{a}\right) \delta\left(k_y + \frac{2\pi}{b}\right)\right. \quad (10.5)$$
$$\left. + \delta\left(k_x + \frac{2\pi}{a}\right) \delta\left(k_y - \frac{2\pi}{b}\right) + \delta\left(k_x + \frac{2\pi}{a}\right) \delta\left(k_y + \frac{2\pi}{b}\right)\right]$$

Problem 2.5. For a harmonic oscillator, the frequency shift near a surface is completely determined by the classical term given by Eq. (2.98). The transition dipole moment components are expressed in terms of the rotationally averaged squared transition dipole moment, μ_{01}^2, as

$$\frac{1}{2}\mu_\|^2 = \mu_\perp^2 = \frac{1}{3}\mu_{01}^2 \quad (10.6)$$

The distance we are searching for can be found as

$$z = \left(\frac{\mu_{01}^2}{6\hbar\gamma} \cdot \frac{n^2 - 1}{n^2 + 1} \right)^{1/3}$$

(10.7)

Substitution of the parameters gives $z \approx 300$ Å.

Problem 2.6.

(a) The hollow adsorption site on the (111) surface of a cubic crystal has the symmetry C_{3v}. The three degrees of freedom of an atom are distributed between the totally symmetric, irreducible representation A_1 and the two-fold representation E, i.e., the total number of different vibrational frequencies is equal to two.

(b) The on-top site has the point symmetry C_{6v}. The corresponding vibrations are classified according to the representations A_1 and E_1, i.e., the number of different frequencies is also two.

(c) The bridge site has the symmetry C_{2v} and the vibrations are classified according to the non-degenerate representations A_1, B_1 and B_2. Therefore the number of different vibrational frequencies is equal to three.

Problem 2.7. The total relaxation rate is found from Eqs (2.140) and (2.141) as

$$R_{tot} = R_{\perp} + R_{\parallel} = \frac{2|\mu_{\perp}^{10}|^2 + |\mu_{\parallel}^{10}|^2}{4\hbar z^3} \, \text{Im} \left(\frac{\tilde{n}^2 - 1}{\tilde{n}^2 + 1} \right)$$

(10.8)

Taking into account the relation (10.6) one finds from here the required distance as

$$z = \left[\frac{|\mu_{01}|^2 \tau}{3\hbar} \, \text{Im} \left(\frac{\tilde{n}^2 - 1}{\tilde{n}^2 + 1} \right) \right]^{1/3}$$

(10.9)

Substitution of the parameters gives $z \approx 210$ Å.

Problem 2.8. The steady-state surface coverage is found from Eq. (2.146) as

$$\theta = \left(1 + \frac{w N_0}{\sigma J} \right)^{-1}$$

(10.10)

with w given by Eq. (2.145). Substituting the parameters, one obtains $\theta \approx 0.05$.

Problem 2.9. The velocity distribution function over the velocity component normal to the surface is given by (see Eq. (2.157))

$$f_\perp(v_z) = \frac{2v_z}{v_T^2} \exp\left(-\frac{v_z^2}{v_T^2}\right), \quad v_z > 0 \qquad (10.11)$$

The maximum is found at $v_z = v_T/\sqrt{2}$, determining the most probable velocity component.

Problem 3.1. Expanding Eq. (3.22) in a power series in ϵ_2'' and using the lowest non-vanishing terms, one finds

$$R_p \approx \frac{(\epsilon_2' - 1)^2 \epsilon_2''^2}{16\epsilon_2'^4} \qquad (10.12)$$

Problem 3.2. Substitution of the quantities $\epsilon_1 = n^2$, $\epsilon_2 = \epsilon_2' + i\epsilon_2''$ and $\epsilon_3 = 1$ into Eqs (3.47) and (3.48), and expansion of the latter to first order in ϵ_2'' results in

$$R_s = 1 - \frac{8\pi d}{\lambda} \cdot \frac{n\epsilon_2''}{n^2 - 1} \cdot \cos\theta_i \qquad (10.13)$$

and

$$R_p = 1 - \frac{8\pi d}{\lambda} \cdot \frac{n\epsilon_2''}{n^2 - 1} \cdot \frac{1 - n^2[1 + (1/\epsilon_2'^2)]\sin^2\theta_i}{1 - (n^2 + 1)\sin^2\theta_i} \cdot \cos\theta_i \qquad (10.14)$$

where we have taken into account that $R_s = R_p = 1$ in the absence of the film.

Problem 3.3. Equation (3.45) determining $r_p(d)$ has a singularity at the Brewster angle and cannot be used directly. The reflection coefficient can be evaluated as follows. Let us introduce a small perturbation in the system by setting $\epsilon_1 = 1 + \alpha$, with α eventually tending to zero. Using Eq. (3.22) for $\theta_i = \arctan(n)$ one finds in the lowest nonvanishing order in α

$$r_p(0) = \frac{\alpha}{4} \cdot \frac{n^2 - 1}{n^2} \qquad (10.15)$$

Multiplying this expression by a power series in α for the quantity $r_p(d)/r_p(0)$, Eq. (3.45), and calculating the limit for $\alpha \to 0$, one obtains

$$r_p(d) = -\frac{\pi i d}{\lambda} \cdot \frac{(\epsilon_2 - 1)(\epsilon_2 - n^2)}{\epsilon_2\sqrt{n^2 + 1}} \qquad (10.16)$$

Problem 3.4. For a macroscopic transition layer we have

$$P_{\alpha j} = \frac{\epsilon_\alpha - 1}{4\pi} E_{\alpha j}, \quad \alpha = 2, 3 \qquad (10.17)$$

The boundary conditions at the interface between the layer and the substrate $(z = d)$ read as

$$E_{2x} = E_{3x} \tag{10.18}$$

$$E_{2y} = E_{3y} \tag{10.19}$$

$$\epsilon_2 E_{2z} = \epsilon_3 E_{3z} \tag{10.20}$$

Assuming that $d \ll \lambda$ and neglecting the variation of $\mathbf{E}_2$ within the layer, one finds

$$\gamma_x = \gamma_y = d\frac{\epsilon_2 - \epsilon_3}{1 - \epsilon_3} \tag{10.21}$$

and

$$\gamma_z = d\frac{\epsilon_2 - \epsilon_3}{\epsilon_2(1 - \epsilon_3)} \tag{10.22}$$

Substitution of the relations (10.21) and (10.22) into Eqs (3.57) and (3.58) gives a result equivalent to Eqs (3.44) and (3.45) $(n_1 = \sqrt{\epsilon_1} = 1)$.

Problem 3.5. For an interface between a metal with the dielectric function ϵ_b and a vacuum, which is described in a three-phase model by the isotropic dielectric function ϵ_2 and effective thickness d, one can write

$$\frac{d\bar{\epsilon}}{dz} = (\epsilon_2 - 1)\delta(z) + (\epsilon_b - \epsilon_2)\delta(z - d) \tag{10.23}$$

where $\bar{\epsilon}$ is the dielectric function of the three-phase system as a whole. The substitution of this equation into Eq. (3.77) gives

$$d_{\|} = \frac{\epsilon_b - \epsilon_2}{\epsilon_b - 1}d \tag{10.24}$$

For the density of the induced charge one can write, analogously to Eq. (3.65),

$$\rho(z) = \frac{1}{4\pi}\{[E_{2z}(z = 0) - E_{1z}(z = 0)]\delta(z) \\ + [E_{3z}(z = d) - E_{2z}(z = d)]\delta(z - d)\} \tag{10.25}$$

where the subscripts 1, 2 and 3 refer to vacuum, transient layer and metal, respectively. Assuming that $d \ll \lambda$ and that, hence, E_{2z} does not change significantly within the layer, the matching conditions at the (12) and (23) interfaces can be written as

$$E_{1z}(z = 0) = \epsilon_2 E_{2z}(z = 0) = \epsilon_2 E_{2z}(z = d) = \epsilon_b E_{3z}(z = d) \tag{10.26}$$

If one takes into account these relations in Eq. (10.25) and substitutes the result into Eq. (3.78), one obtains

$$d_{\perp} = \frac{\epsilon_b - \epsilon_2}{\epsilon_2(\epsilon_b - 1)}d \tag{10.27}$$

Finally, substitution of Eqs (10.24) and (10.27) along with Eqs (3.5), (3.6) and (3.7) into Eqs (3.75) and (3.76) gives, after some algebraic manipulation, the required result with $\epsilon_1 = 1$ and $\epsilon_3 = \epsilon_b$.

Problem 3.6. A TE wave has electric and magnetic fields represented by the vectors $(0, E_y, 0)$ and $(H_x, 0, H_z)$, respectively, where each component has a form similar to Eq. (3.81) for $z < 0$ and to Eq. (3.82) for $z > 0$. Their substitution into Maxwell's equation $\nabla \times \mathbf{E} = -(1/c)(\partial \mathbf{H}/\partial t)$ gives the following two equations:

$$\kappa_1 E_{y1} = i\frac{\omega}{c} H_{x1}, \qquad z < 0 \tag{10.28}$$

$$\kappa_2 E_{y2} = -i\frac{\omega}{c} H_{x2}, \qquad z > 0 \tag{10.29}$$

The tangential components of both electric and magnetic field vectors must be continuous across the boundary $z = 0$. This leads to the equation $\kappa_1 = -\kappa_2$ which cannot be fulfilled because both κ_1 and κ_2 are positive.

Problem 3.7. Substituting $\epsilon_1 = 1$ and $\epsilon_2 = 1 - (\lambda^2/\lambda_p^2)$ into Eqs (3.91) and (3.92), one obtains for the penetration depth into vacuum

$$\delta_1 \approx \frac{\lambda^2}{2\pi\lambda_p} \tag{10.30}$$

and for the penetration depth into metal

$$\delta_2 \approx \frac{\lambda_p}{2\pi} \tag{10.31}$$

where we have taken into account that $\lambda \gg \lambda_p$. For the given parameters we find $\delta_1 \approx 6900$ Å and $\delta_2 \approx 130$ Å.

Problem 3.8. The dispersion relation for a surface polariton, Eq. (3.89), with $\epsilon_i = 1 - (\omega_{pi}^2/\omega^2)$, $i = 1, 2$, can be reduced to the form

$$q = \frac{1}{c}\sqrt{\frac{(\omega_{p1}^2 - \omega^2)(\omega_{p2}^2 - \omega^2)}{2\omega^2 - \omega_{p1}^2 - \omega_{p2}^2}} \tag{10.32}$$

The surface plasmon polariton can exist if the quantity under the square-root is positive. This occurs in the frequency ranges $\omega_{p1} < \omega < [(\omega_{p1}^2 + \omega_{p2}^2)/2]^{1/2}$ and $\omega > \omega_{p2}$ provided that $\omega_{p1} < \omega_{p2}$. However, in the latter range both ϵ_1 and ϵ_2 are positive, thus contradicting Eq. (3.88) from which the dispersion relation follows. Therefore, a surface plasmon polariton can be excited only in the former range. In the nonretarded limit $c \to \infty$ this corresponds to

$$\omega_s = \sqrt{\frac{\omega_{p1}^2 + \omega_{p2}^2}{2}} \tag{10.33}$$

Problem 3.9. The dielectric function of an n-type semiconductor is determined as

$$\epsilon(\omega) = \epsilon_\infty + (\epsilon_0 - \epsilon_\infty)\frac{\omega_{TO}^2}{\omega_{TO}^2 - \omega^2} - \frac{\omega_p^2}{\omega^2} \tag{10.34}$$

where ω_p is given by Eq. (2.65).

A general view of the function $\epsilon(\omega)$ is shown in Fig. 10.2a. The variation of the dielectric function for a larger scale in the regions where it intersects the line $\epsilon = -1$ is shown in Figs 10.2b and c. Surface polaritons can exist at a semiconductor – vacuum interface in the frequency range where $\epsilon(\omega) < -1$. This occurs for $\omega < 160\ \mathrm{cm}^{-1}$ and for $180\ \mathrm{cm}^{-1} < \omega < 640\ \mathrm{cm}^{-1}$.

Problem 3.10. According to Eq. (3.107), the condition for surface polariton excitation reads as

$$\sin\theta_i = \sqrt{\frac{\epsilon(\lambda)}{\epsilon(\lambda) + 1}} + m\frac{\lambda}{d} \tag{10.35}$$

with $m = 0, \pm1, \pm2,\ldots$ Substitution of the parameters gives the equation

$$\sin\theta_i = \frac{3}{2\sqrt{2}} + \frac{m}{3} \tag{10.36}$$

The modulus of the right-hand side of this relation must not exceed unity. This restricts the possible values of m to $m = -1, -2, -3$. The corresponding angles are equal to $\theta_{-1} = 46.7°$, $\theta_{-2} = 23.2°$ and $\theta_{-3} = 3.5°$.

Problem 3.11. The surface polariton propagation length is determined as $L = [2\,\mathrm{Im}(q)]^{-1}$ with q the SP wave vector. On the other hand, the ATR minimum reaches zero if $\Gamma_i = \mathrm{Im}(q)$ equals Γ_{rad} (see Eq. (3.111)). Under such conditions the full width at half reflection minimum is given by $4\Gamma_i$. This quantity can be alternatively found as $(2\pi n/\lambda)\cos\theta_m\Delta\theta$. Finally, one obtains

$$L = \frac{1}{2\Gamma_i} = \frac{\lambda}{\pi n \cos\theta_m \Delta\theta} \tag{10.37}$$

Substitution of the parameters gives $L \approx 20\ \mu\mathrm{m}$.

Problem 4.1. Use the solution of Problem 3.3. Taking into account that $n \ll \sqrt{|\epsilon_2'|}$, one finds

$$r_p(d) \approx -\frac{i\pi d}{\lambda} \cdot \frac{\epsilon_2}{\sqrt{n^2 + 1}} \tag{10.38}$$

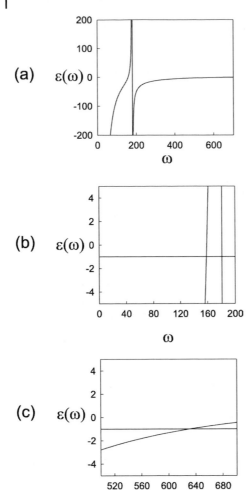

Fig. 10.2 (a) Plot of $\epsilon(\omega)$ calculated for the given parameters. (b) and (c) showing ranges of ω (in cm^{-1}) which contain the points of intersection with the line $\epsilon = -1$.

The leading term in d is found for r_s in zeroth order, i.e., $r_s(0)$. Substitution for the angle of incidence $\theta_i = \theta_B = \arctan(n)$ gives

$$r_s(0) = \frac{1 - n^2}{1 + n^2} \tag{10.39}$$

Hence for $\rho = r_p/r_s$

$$\rho \approx -\frac{i\pi d}{\lambda} \cdot \frac{\sqrt{1 + n^2}}{1 - n^2}(\epsilon_2' + i\epsilon_2'') \tag{10.40}$$

Separating real and imaginary parts, one finally gets

$$\Psi \approx \arctan\left(-\frac{\pi d}{\lambda} \cdot \frac{\sqrt{1+n^2}}{1-n^2}\sqrt{\epsilon_2'^2 + \epsilon_2''^2}\right) \tag{10.41}$$

and

$$\Delta \approx \arctan\left(-\frac{\epsilon_2'}{\epsilon_2''}\right) \tag{10.42}$$

Problem 4.2. The only normal mode displayed in the IRAS spectrum from W(100)−H is the vibration normal to the surface. The intensity of the spectral lines corresponding to the other two vibrations parallel to the metal surface are suppressed because of the pseudo-selection rule.

Problem 4.3. One of the lines corresponds to the vibration of H atoms perpendicular to the surface, whereas the other one originates from the two-fold degenerate vibration parallel to the surface. In the spectrum excited by s-polarized light one will observe only the latter line. Since the frequency of vibrations parallel to the surface is degenerate it can be due to adsorption either at a hollow or at an on-top site (symmetry C_{4v}).

Problem 4.4. The surface polariton wave vector has the form

$$q = \frac{2\pi}{\lambda}\sqrt{\frac{\epsilon' + i\epsilon''}{1 + \epsilon' + i\epsilon''}} \tag{10.43}$$

where $\epsilon = \epsilon' + i\epsilon''$ is the dielectric function of a crystal. If $|\epsilon| \gg 1$, then in the lowest nonvanishing orders in $1/|\epsilon|$ one has

$$\mathrm{Re}(q) = \frac{2\pi}{\lambda} \tag{10.44}$$

and

$$\mathrm{Im}(q) = \frac{\pi}{\lambda} \cdot \frac{\epsilon''}{\epsilon'^2 + \epsilon''^2} \tag{10.45}$$

In this case $\mathrm{Im}(q) \ll \mathrm{Re}(q)$, implying the inequality $L \gg \lambda$. The corresponding propagation length is found as

$$L = \frac{\lambda}{2\pi} \cdot \frac{\epsilon'^2 + \epsilon''^2}{\epsilon''} \tag{10.46}$$

Problem 4.5. At frequencies $\omega \approx \omega_{TO}$ one has (see Eq. (4.7)) $\epsilon' \approx \epsilon_\infty$ and

$$\epsilon'' \approx -\frac{(\epsilon_0 - \epsilon_\infty)\omega_{TO}}{2\Gamma} \tag{10.47}$$

These relations imply that the inequalities $\epsilon' \ll \epsilon''$ and $|\epsilon''| \gg 1$ are fulfilled. Then one can use the solution of Problem 4.4 and thus obtain

$$L \approx \frac{c(\epsilon_0 - \epsilon_\infty)}{2\Gamma} \tag{10.48}$$

Problem 5.1. The ratio of the reflection coefficients for p- and s-polarizations can be represented as

$$\frac{|r_p| \exp(i\delta_p)}{|r_s| \exp(i\delta_s)} = \tan \Psi \exp(i\Delta) \tag{10.49}$$

It follows that $\Delta = \delta_p - \delta_s$. Taking into account that the reflection was measured at the Brewster angle, we can conclude that the main contribution to the variation of Δ during adsorption originates from the quantity r_p, i.e., $\delta_s \approx const$. From Fig. 5.2 the variation of δ_p is calculated as $\arccos(-0.36) - \arccos(-0.15) \approx 13°$.

Problem 5.2. A semiconductor is transparent at energies below the forbidden gap and hence its dielectric function is real, i.e., $\epsilon_3 = n_3^2$. Substituting this quantity along with $\epsilon_1 = n_1^2 = 1$ and $\epsilon_2 = (n_2 + i\kappa_2)^2$ into Eq. (3.49), one obtains

$$\frac{\Delta R}{R} = \frac{4n_2}{n_3^2 - 1}(\alpha d) \tag{10.50}$$

where

$$\alpha = \frac{4\pi\kappa_2}{\lambda} \tag{10.51}$$

is the absorption coefficient of the film.

Problem 5.3. At normal incidence Eq. (3.51) is reduced to:

$$\frac{r_{pp}(d)}{r_{pp}(0)} = 1 + \frac{4\pi id}{\lambda} \cdot \frac{\epsilon_b - \bar{\epsilon} - \Delta\epsilon \cos 2\psi}{\epsilon_b - 1} \tag{10.52}$$

Then one obtains for the RAS signal in the lowest nonvanishing order in d

$$\frac{\Delta r}{r} \approx \frac{r_{pp}(d, \psi = 90°) - r_{pp}(d, \psi = 0°)}{r_{pp}(0)} = \frac{4\pi id}{\lambda} \cdot \frac{\epsilon_{yy} - \epsilon_{xx}}{\epsilon_b - 1} \tag{10.53}$$

where we have used the definition of $\Delta\epsilon$ from the text.

Problem 5.4. According to the Franck–Condon principle, the maximum absorption of Na adatoms corresponds to a vertical transition from the minimum of the ground electronic state to the excited electronic state (see Fig. 5.8). This condition can be written in terms of the Na adatom binding energies in the ground and excited states, E_{bg} and E_{be}, as

$$E_{bg} + \frac{hc}{\lambda_0} - E_{be} \leq \frac{hc}{\lambda_s} \tag{10.54}$$

where λ_0 is the wavelength of the transition in Na atoms far from the surface and λ_s is the wavelength corresponding to the absorption band maximum of Na adatoms. As seen from Fig. 5.7, $\lambda_s > \lambda_0$. From here we conclude that

$$E_{be} - E_{bg} \geq hc \left(\frac{1}{\lambda_0} - \frac{1}{\lambda_s} \right) > 0 \tag{10.55}$$

i.e., the adsorption energy in the $3P$ state is larger than the adsorption energy in the $3S$ state.

Problem 5.5. The minima and maxima in Fig. 5.10 originate from interference between the radiation emitted by an atomic dipole and the part reflected from the silver surface. This becomes evident if the distance between the adjacent minima (or maxima), Δz, obeys the relation $2(\Delta z)n = \lambda$. Substitution of the parameters gives $\Delta z = 2040$ Å, being approximately fulfilled as one can see in Fig. 5.10.

Problem 5.6. The dielectric function satisfies the condition $\epsilon'' \ll |\epsilon'|$ and hence one can use Eq. (3.122) for the maximum field enhancement $\eta(\lambda)$ with $A = 1/3$ for a sphere. Under the assumption that the dielectric function does not vary essentially over the Raman shift, the overall SERS enhancement can be estimated as $\eta^2(\lambda)$. Substitution of the parameters gives $\eta^2 \approx 2.1 \times 10^9$ for a sphere and $\eta^2 \approx 2.6 \times 10^{11}$ for a spheroid.

Problem 6.1. Buried layers in a centrosymmetric medium can be investigated by second harmonic spectroscopy. Electron energy loss techniques cannot be used since they suffer from a small escape depth of the generated electrons.

Problem 6.2. The atomic displacement of a dipole parallel to a metallic surface is strongly suppressed. Thus one would not expect to observe frequency doubled radiation from a metallic surface that is irradiated by s-polarized light.

Problem 6.3. One would expect to see a very weak SHG intensity at both grazing and normal incidence and a maximum in between, the position of which is determined by the relevant Fresnel factors.

Problem 6.4. From Eq. (6.19) one calculates again of a factor of 10^5.

Problem 6.5. The second-order nonlinear optical signal from a metal surface is determined by Friedel oscillations, which have a typical wavelength of $\lambda_F = \pi/k_F$ with k_F the Fermi wavevector. One finds typical wavelengths of the order of 0.5 to 1 nm.

Problem 7.1. The Doppler width is determined as

$$\Delta\omega_D = 4\sqrt{\ln 2}\,\frac{\pi v_T}{\lambda} \tag{10.56}$$

with v_T the most probable thermal velocity, Eq. (2.153). Thus $\Delta\omega_D \approx 2.8\,\text{GHz}$, much larger than the natural linewidth. Therefore the required condition is fulfilled if the homogeneous broadening is determined mainly by collisional broadening, i.e., $\gamma_c \approx \Delta\omega_D$. Then the number density of Cs atoms is found to be $n \approx 3.7 \cdot 10^{15}\,\text{cm}^{-3}$, corresponding to a pressure of $P \approx 0.16$ Torr.

Problem 7.2. The transient term responsible for the spectral narrowing cannot be neglected if, according to Eq. (7.29),

$$\frac{2\pi v_T}{\lambda}\cos\theta_t \geq \gamma \tag{10.57}$$

where $\gamma = \gamma_n + \gamma_c$ and the angle θ_t is related to the incidence angle through Snell's law. From here one finds the required angle as

$$\theta_i = \arcsin\left[\frac{1}{n_w}\sqrt{1 - \left(\frac{\lambda\gamma}{2\pi v_T}\right)^2}\right] \tag{10.58}$$

The calculation gives $\theta_i \approx 34°$. This value is below the critical angle $\theta_c = 41.8°$.

Problem 7.3. The fluorescence spectrum will be sensitive to the population of the scattered atoms if the evanescent wave penetration depth, δ, is less than, or comparable to, the population memory length l_1 (see Section 2.4.3). This condition is fulfilled at incidence angles above θ_i satisfying the equation

$$\frac{v_T}{\gamma_n} = \frac{\lambda}{2\pi\sqrt{n_p^2\sin^2\theta_i - 1}} \tag{10.59}$$

where we have taken into account that $T_1 = 1/\gamma_n$. From here one finds

$$\theta_i = \arcsin\left[\frac{1}{n_p}\sqrt{1 + \left(\frac{\lambda\gamma_n}{2\pi v_T}\right)^2}\right] \tag{10.60}$$

Substitution of the parameters gives $\theta_i \approx 41.82°$, slightly above the critical angle $\theta_c = 41.81°$.

Problem 7.4. The vapor spectra excited by an evanescent wave will depend strongly on the change of the vapor polarization in vapor–surface scattering if the evanescent wave penetration depth is less than or comparable with the polarization memory length, i.e.,

$$\delta \leq \frac{2v_T}{\gamma_n + \gamma_c} \tag{10.61}$$

The minimal penetration depth which occurs at grazing incidence ($\theta_i \approx 90°$) is found as

$$\delta_{\min} = \frac{\lambda}{2\pi\sqrt{n_p^2 - 1}} \tag{10.62}$$

From here one obtains the maximum collisional broadening corresponding to the condition (10.61) as

$$\gamma_{c,max} = \frac{4\pi v_T \sqrt{n_p^2 - 1}}{\lambda} - \gamma_n \tag{10.63}$$

The calculation gives $\gamma_{c,max} = 1.1 \times 10^{10}$ s^{-1}. Accordingly, $n = 2.5 \times 10^{16}$ cm^{-3} and the vapor pressure $P = 0.78$ Torr.

Problem 7.5. The velocity distribution functions over the velocity components parallel and perpendicular to the surface are given by

$$f_\|(v_x) = \frac{1}{\sqrt{\pi}v_T} \exp\left(-\frac{v_x^2}{v_T^2}\right) \tag{10.64}$$

and

$$f_\perp(v_z) = \frac{2v_z}{v_T^2} \exp\left(-\frac{v_z^2}{v_T^2}\right), \quad v_z > 0 \tag{10.65}$$

respectively, with v_T the most probable thermal velocity, Eq. (2.153). The maximum of the function $f_\|$ occurs at $v_x = 0$, whereas the maximum of the function $f_\perp$ takes place at $v_z = v_T/\sqrt{2}$ (see Problem 2.9). Thus the frequency shift between the two maxima, $\omega_\perp$ and $\omega_\|$, in the fluorescence spectra is determined as

$$\Delta\omega = \omega_\perp - \omega_\| = \frac{1}{\sqrt{2}}kv_T \tag{10.66}$$

This shift is positive if the light beam normal to the surface propagates from the substrate side and it is negative if the beam propagates in the reverse direction. Substituting the parameters, one obtains $\Delta\omega = 3.9$ GHz or $\Delta\nu = 620$ MHz.

Problem 8.1. The minimum, diffraction-limited spot size obtainable with a given lens can be achieved with spatially coherent light. This light is provided, for example, by a laser or by inserting spatial filters in the light beam.

Problem 8.2. Use the definition of the "resolving power" and $\tan\theta \approx \theta$ to find that the minimum area is roughly the square of the applied wavelength.

Problem 8.3. If one intends to obtain resolution better than the wavelength of the imaging light one has to implement the phase of the electromagnetic wave. This is done, for example, in the interference microscope. One obtains resolution below one nanometer along the normal to the surface.

Problem 8.4. In order to determine an adsorbate thickness from an ellipsometric measurement one needs to know the dielectric function of the investigated system. This function depends on temperature and thus can give rise to a large uncertainty in the measurement.

Problem 8.5. Since the SH intensity is proportional to the square of the transition dipole moment one only has to invoke that the transition dipole moment for s-polarized excitation radiation is proportional to $\sin^2\theta$ in order to find the required equation.

Problem 9.1. The characteristic size of the object d and the wavelength λ have to be compared. Depending on whether λ is smaller or larger then d, one deals with far-field vs. near-field effects. In the far field the divergence angle is constant (both incident and diffracted waves are effectively plane) and the relevant diffraction theory is Fraunhofer diffraction. In the near field the wave fronts are curved and one has to apply Fresnel diffraction theory.

Problem 9.2. The minimum size of the imaging aperture for a SNOM is given by fundamental and technical limitations. The penetration depth of the light into the aluminum cover (the skin depth) represents a fundamental limit of the order of 10 to 30 nm. Technical limitations are given by the inhomogeneity (grain formation) of the metal coating and by the necessity of having a certain acceptance angle so as to obtain a measurable signal intensity.

Problem 9.3. Since a SNOM works by the same principle as an AFM, the topographical resolution is given by the effective fiber size. Due to the fiber etching procedure, multicrystallinity of the glass and also due to the rough metal coating this effective size can be rather large. The effective SNOM tip is then reflected in the topographic shapes that are imaged.

Problem 9.4. Surface-enhanced Raman scattering is driven mainly by chemical and electromagnetic effects. If one neglects the chemical effects one obtains a large Raman scattering intensity if the electromagnetic field enhancement is strong. The field, for example, for a sphere is inversely proportional to the radius of the sphere. Thus one obtains high field strengths at sharp tips.

Glossary

Adatom   An adsorbed atom.

Adlayer   An adsorbed layer.

Admolecule   An adsorbed molecule.

Adsorbate   The material in the adsorbed state.

Adsorbent   The substrate on which adsorption occurs.

Adsorption   Bonding of foreign atoms or molecules on a crystal surface.

Adsorption isotherm   Dependence of surface coverage on the pressure of the surrounding gas phase at a constant temperature under thermodynamical equilibrium between a surface and a gas.

Adsorption potential   The interaction potential between an atom or a molecule and a crystal surface.

Allowed energy band   An energy band consisting of electronic states available for electrons in the bulk crystal.

Associative desorption   Desorption which follows the formation of a molecule from separately adsorbed species.

Attenuated total reflection (ATR)   Attenuation of the reflected wave intensity in total internal reflection due to energy losses in the medium which covers the face where reflection occurs. This phenomenon is used to monitor the excitation of surface polaritons in Otto and Kretschmann configurations.

Optics and Spectroscopy at Surfaces and Interfaces. Vladimir G. Bordo and Horst-Günter Rubahn
Copyright © 2005 WILEY-VCH Verlag GmbH & Co. KGaA, Weinheim
ISBN: 3-527-40560-7

Bravais lattice   An infinite set of points generated by a set of all possible translations determined by the primitive lattice vectors.

Brewster angle   The angle of incidence for which the reflection coefficient for a p-polarized electromagnetic wave is equal to zero.

Chemisorption   Chemical adsorption.   Adsorption in which the forces involved are valence forces. It is accompanied by the formation of a chemical bond between a foreign atom or molecule and the surface atoms of a crystal.

Cosine law   An assumption that the intensity of an atomic flux desorbing from a surface under nonequilibrium conditions varies as $\cos\theta$, with θ the angle between the atomic velocity vector and the surface normal.

Critical angle   Angle of incidence above which total internal reflection occurs.

Cyclic voltammogram   A current–voltage curve obtained when the potential applied to the electrode–electrolyte interface is ramped to some switch potential and then is scanned back to its initial value.

Desorption   The opposite process to adsorption, i.e., the breaking of the bonding between adsorbed atoms or molecules and a crystal surface.

Dispersion relation   A functional relationship between frequency and wave vector (energy and momentum) of a harmonic wave.

Electrical double layer   An interface dipole layer between a metal surface and an electrolyte.

Electronic surface states   Electronic states corresponding to electron wave functions localized near a surface.

Evanescent wave   An electromagnetic wave which propagates along the interface in total internal reflection. Its amplitude decays exponentially into the medium with the smaller index of refraction.

Forbidden energy gap   An energy gap between two allowed energy bands. One distinguishes a direct gap in which the electron wave function has s-like character at the bottom and p-like character at the top, and an inverted gap in which the order of the wave functions is inverted in comparison with the s- and p-levels of isolated atoms.

First Brillouin zone   The primitive cell in the reciprocal lattice containing the point **k=0**. All physically different electronic states in a crystal can be characterized by a wave vector reduced to the first Brillouin zone.

HOMO   Highest Occupied Molecular Orbital. The molecular orbital of an adsorbed molecule which is nearest to the Fermi level of the metal substrate among the ones located below it.

Image states   Electronic states bound to a metal surface which occur in the Coulomb potential of the electron image. They form a series of energy levels similar to that of the hydrogen-like atoms.

Infrared region   The spectral range between the red edge of visible light ($\lambda \sim 0.76$ μm) and the short radio waves ($\lambda \approx 1$–2 mm).

Knudsen layer   A gas layer bordering a solid surface where interatomic collisions are negligible. The thickness is of the order of the mean free path.

Kretschmann configuration (geometry)   An ATR configuration represented by a sample film deposited onto a prism surface.

Langmuir isotherm   The adsorption isotherm which is obtained in the Langmuir model of adsorption.

Langmuir model of adsorption   A model of adsorption in which one assumes that adsorption can occur only at free adsorption sites.

Langmuir–Blodgett film   A set of monolayers of organic molecules deposited onto a solid substrate. The structure of such a film can be controlled at the molecular level.

LUMO   Lowest Unoccupied Molecular Orbital. The molecular orbital of an adsorbed molecule which is nearest to the Fermi level of the metal substrate among the ones located above it.

Monolayer   A one-atom or one-molecule thick adsorbed layer.

Nonlocal optical response   Optical response of a medium to an external electromagnetic field where the induced polarization at a given point is related to the electric field at neighboring points.

Optical region   The spectral range including near-infrared, visible and near-ultraviolet light.

Otto configuration (geometry)   An ATR configuration represented by a prism suspended above the sample with the help of a distance holder.

p-polarization   The polarization of an electromagnetic wave incident on the interface at which the electric field vector lies in the plane of incidence.

Physisorption   Physical adsorption. Adsorption in which the forces involved are intermolecular (van der Waals) forces. It is not accompanied by a significant change in the electronic orbitals of the species involved.

Potential of zero charge (PZC)   The value of the potential applied to the electrode–electrolyte interface at which the electrode is uncharged.

Pseudo-selection rules   Attenuation of spectral lines of molecules adsorbed on metal surfaces which correspond to vibrational modes accompanied by the atomic displacements parallel to the surface.

Reciprocal lattice   A Bravais lattice in the space of wave vectors.

Reconstruction   A modification of the surface structure occurring upon a crystal cleavage when the surface atoms are shifted along the surface with respect to their positions in the bulk.

Reflection coefficient   The ratio of the reflected to the incident electric field amplitude.

Reflectivity   The ratio of the reflected to the incident time-averaged Poynting vector component normal to the interface.

Relaxation of a surface   A modification of distances between the atomic planes near to the surface, which occurs upon crystal cleavage.

Resolving power   Measure of the ability of an optical system to form distinct images of two objects.

s-polarization   The polarization of an electromagnetic wave incident on the interface at which the electric field vector is perpendicular to the plane of incidence.

Scattering kernel   The kernel in the integral equation which relates the velocity distribution function of the gas atoms scattered by the surface to that of the atoms arriving at the surface.

Selective reflection   Reflection from an interface between a transparent dielectric and an atomic vapor using light of a frequency close to the transition frequency of vapor atoms.

Selvedge region   The region at the edge of a crystal.

Shockley states   Electronic surface states which are obtained in a model implementing an inverted forbidden energy gap.

Sticking probability   The probability that an atom or a molecule approaching the surface will be adsorbed on it.

Superlattice   A surface Bravais lattice characterizing a surface reconstruction or an ordered arrangement of adsorbed atoms.

Surface phonon   A collective vibration of atoms localized in the surface region of a crystal.

Surface plasmon   A collective excitation of conduction electrons localized near a conductor surface.

Surface polariton   A coupled electromagnetic mode arising from interaction between incident light and an excitation in a crystal which propagates along the crystal surface. One distinguishes between surface plasmon polaritons, surface phonon polaritons, etc., originating from the interaction between light and plasmons, phonons and other crystal excitations.

Surface resonances   Electronic surface states which are degenerate in energy with the bulk states at some wave vectors. An electron in such a state can propagate from the surface into the bulk crystal.

Surface-enhanced Raman scattering (SERS)   Raman scattering from molecules adsorbed on rough metal surfaces which give rise to huge cross-section enhancements.

Tamm states   Electronic surface states. Sometimes one refers to Tamm states as those electronic surface states which are obtained in the model of tightly bound electrons. Such states correspond to the case of a direct forbidden energy gap.

Total internal reflection   Reflection that occurs when a wave is incident from a medium with a large index of refraction to a low-index medium, at angles of incidence greater than the critical angle.

Transmission coefficient   The ratio of transmitted to incident electric field amplitude.

Transmissivity   The ratio of the transmitted to the incident time-averaged Poynting vector component normal to the interface.

Unit cell   A spatial arrangement of atoms which is the basis for a complete description of the crystal structure.

van der Waals potential   The interaction potential between instantaneous dipoles of the species involved, originating from the quantum-mechanical fluctuations of their electron clouds. It describes the interaction between an atom or a molecule and a crystal surface at large distances apart.

Bibliography

Abbe, E. (1873). Beiträge zur Theorie des Mikroskops und der mikroskopischen Wahrnehmung. *Arch. Mikrosk. Anatomie*, 9:413.

Abelès, F. and Lopez-Rios, T. (1982). Surface polaritons at metal surfaces and interfaces. In Agranovich, V. and Mills, D., editors, *Surface Polaritons*, pages 239 – 274. North-Holland Publishing Company.

Akul'shin, A., Velichanskii, V., Zibrov, A., Nikitin, V., Sautenkov, V., Yurkin, E., and Senkov, N. (1982). Collisional broadening of intra-Doppler resonances of selective reflection on the D_2 line of cesium. *JETP Lett.*, 36:303 – 307.

Andrews, D. and Hands, I. (1996). Second harmonic generation in partially ordered media and at interfaces: Analysis of dynamical and orientational factors. *Chem.Phys.*, 213:277 – 294.

Andrieu, S. and d'Avitaya, F. (1991). Ga adsorption on Si(111) analyzed by RHEED and *in situ* ellipsometry. *J. Cryst. Growth*, 112:146 – 152.

Ash, E. and Nicholls, G. (1972). Super-resolution aperture scanning microscope. *Nature*, 237:510 – 512.

Aussenegg, F., Leitner, A., and Gold, H. (1995). Optical second-harmonic generation of metal-island films. *Appl. Phys. A*, 60:97 – 101.

Bagatolli, L. (2005). private communication.

Balzer, F., Bammel, K., and Rubahn, H.-G. (1993). Laser investigation of Na atoms deposited via inert spacer layers close to metal surfaces. *J. Chem. Phys.*, 98:7625 – 7635.

Balzer, F. and Rubahn, H.-G. (1995). Third-order nonlinear optics of Na clusters bound to dielectric surfaces. *Chem. Phys. Lett.*, 238:77 – 81.

Balzer, F. and Rubahn, H.-G. (1998). Nonlinear optics of rough cluster films. *Proc. SPIE*, 3272:23 – 34.

Beckerle, J., Casassa, M., Cavanagh, R., Heilweil, E., and Stephenson, J. (1990). Ultrafast infrared response of adsorbates on metal surfaces: Vibrational lifetime of CO/Pt(111). *Phys. Rev. Lett.*, 64:2090 – 2093.

Beermann, J., Bozhevolnyi, S., Balzer, F., and Rubahn, H.-G. (2005). Two-photon near-field mapping of molecular orientations in organic nanofibers. *Laser Physics Letters*, submitted.

Beermann, J., Bozhevolnyi, S., Bordo, V., and Rubahn, H.-G. (2004). Two-photon mapping of local molecular orientations in hexaphenyl nanofibers. *Opt.Commun.*, 237:423 – 429.

Berkovic, G. (1995). Strategies for measuring third harmonic generation from ultrathin films. *Chem. Phys. Lett.*, 241:355 – 359.

Besenbacher, F. (1996). Scanning tunneling microscopy studies of metal surfaces. *Rep. Prog. Phys.*, 59:1737 – 1802.

Binnig, G., Quate, C., and Gerber, C. (1986). Atomic force microscope. *Phys. Rev. Lett.*, 56:930 – 933.

Binnig, G., Rohrer, H., Gerber, C., and Weibel, E. (1982). Surface studies by scanning tunneling microscopy. *Phys. Rev. Lett.*, 49:57 – 61.

Bloembergen, N., Chang, R., Jha, S., and Lee, C. (1968). Optical second-harmonic generation in reflection from media with inversion symmetry. *Phys. Rev.*, 174:813 – 822.

Bonch-Bruevich, A., Maksimov, Y., and Khromov, V. (1985). Variation of the absorption spectrum of sodium atoms when they are adsorbed on a sapphire surface. *Optics Spectrosc. (USSR)*, 58:854 – 856.

Bordo, V., Henkel, C., Lindinger, A., and Rubahn, H.-G. (1997). Evanescent wave fluorescence spectra of Na atoms. *Opt. Comm.*, 137:249 – 253.

Bordo, V., Loerke, J., and Rubahn, H.-G. (2001). Two-photon evanescent wave spectroscopy: A new account to gas-solid dynamics in the boundary layer. *Phys. Rev. Lett.*, 86:1490 – 1493.

Bordo, V., Marowsky, G., Zhang, J., and Rubahn, H.-G. (2005). Resonant effects in optical second-harmonic generation from alkali covered Si(111)7×7. *Europhys.Lett.*, 69:61 – 67.

Bordo, V. and Rubahn, H.-G. (1999a). Laser-controlled adsorption of Na atoms in evanescent wave spectroscopy. *Opt. Express*, 4:59 – 66.

Bordo, V. and Rubahn, H.-G. (1999b). Two-photon evanescent-wave spectroscopy of alkali-metal atoms. *Phys. Rev. A*, 60:1538 – 1548.

Bordo, V. and Rubahn, H.-G. (2003). Determination of the gas-surface scattering kernel from two-photon evanescent-volume-wave fluorescence spectra. *Phys. Rev. A*, 67:012901–1 – 8.

Bordo, V. and Rubahn, H.-G. (2004). Evanescent wave spectroscopy. In Nalwa, H., editor, *Encyclopedia of Nanoscience and Nanotechnology*, volume 3, pages 287 – 295. American Scientific Publishers.

Born, M. and Wolf, E. (1975). *Principles of Optics*. Pergamon Press, Oxford.

Boyd, G., Rasing, T., Leite, J., and Shen, Y. (1984). Local-field enhancement on rough surfaces of metals, semimetals, and semiconductors with the use of optical second-harmonic generation. *Phys. Rev. B*, 30:519 – 526.

Boyd, G., Shen, Y., and Hänsch, T. (1986). Continuous-wave second-harmonic generation as a surface microprobe. *Opt. Lett.*, 11:97 – 99.

Brako, R. and Newns, D. (1982). Differential cross section for atoms inelastically scattered from surfaces. *Phys. Rev. Lett.*, 48:1859 – 1862.

Bratz, A. and Marowsky, G. (1990). Angular dependence of surface second harmonic generation. *Molecular Engineering*, 1:59 – 65.

Brevet, P.-F. (1997). *Surface Second Harmonic Generation*. Presses Polytechniques, Lausanne.

Brewer, J. and Rubahn, H.-G. (2005). private communication.

Brodsky, A., Daikhin, L., and Urbakh, M. (1985). Electrochemical optical spectroscopy of metals. *Uspekhy Khimie*, 54:3 – 32.

Brusdeylins, G., Rechsteiner, R., Skofronick, J., Toennies, J., Benedek, G., and Miglio, L. (1985). Observation of surface optical phonons in NaF(001) by inelastic He-atom scattering. *Phys. Rev. Lett.*, 54:466 – 469.

Burgmans, A., Schuurmans, M., and Bölger, B. (1977). Transient behavior of optically excited vapor atoms near a solid interface as observed in evanescent wave emission. *Phys. Rev. A*, 16:2002 – 2007.

Butcher, P. and Cotter, D. (1990). *The Elements of Nonlinear Optics*. Cambridge University Press, Cambridge.

Byers, J., Yee, H., Petralli-Mallow, T., and Hicks, J. (1994). Second harmonic generation-circular dichroism spectroscopy from chiral monolayers. *Phys.Rev.B*, 39:14643 – 14647.

Campion, A. and Kambhampati, P. (1998). Surface-enhanced Raman scattering. *Chem. Soc. Rev.*, 27:241 – 250.

Cavanagh, R., Heilweil, E., and Stephenson, J. (1994). Time-resolved measurement of energy transfer at surfaces. *Surf. Sci.*, 299/300:643 – 655.

Cercignani, C. (2000). *Rarefied Gas Dynamics. From Basic Concepts to Actual Calculations.* Cambridge University Press, Cambridge.

Chance, R., Prock, A., and Silbey, R. (1975a). Comments on the classical theory of energy transfer. *J. Chem. Phys.*, 62:2245 – 2253.

Chance, R., Prock, A., and Silbey, R. (1975b). Frequency shifts of an electric-dipole transition near a partially reflecting surface. *Phys. Rev. A*, 12:1448 – 1452.

Chekalin, S., Letokhov, V., Likhachev, V., and Movshev, V. (1984). Laser photoion projector. *Appl. Phys. B*, 33:57 – 61.

Chemla, D., Heritage, J., Liao, P., and Isaacs, E. (1983). Enhanced four-wave mixing from silver particles. *Phys. Rev. B*, 27:4553 – 4558.

Chen, C. (1993). *Introduction to Scanning Tunneling Microscopy*. Oxford University Press.

Chen, C., de Castro, A., and Shen, Y. (1981). Surface-enhanced second-harmonic generation. *Phys. Rev. Lett.*, 46:145 – 148.

Chen, C., de Castro, A., Shen, Y., and DeMartini, F. (1979). Surface coherent anti-Stokes Raman spectroscopy. *Phys. Rev. Lett.*, 43:946 – 949.

Chen, C., Heinz, T., Ricard, D., and Shen, Y. (1983). Surface-enhanced second-harmonic generation and Raman scattering. *Phys. Rev. B*, 27:1965 – 1979.

Chen, J., Bower, J., Wang, C., and Lee, C. (1973). Optical second-harmonic generation from submonolayer Na-covered Ge surfaces. *Opt. Comm.*, 9:132 – 134.

Chiarotti, G., Nannarone, S., Pastore, R., and Chiaradia, P. (1971). Optical absorption of surface states in ultrahigh vacuum cleaved (111) surfaces of Ge and Si. *Phys. Rev. B*, 4:3398 – 3402.

Cojan, J.-L. (1954). Contribution a l'étude de la réflexion sélective sur les vapeurs de mercure de la radiation de résonance du mercure. *Ann. Phys. (Paris)*, 9:385 – 440.

Comsa, G. and David, R. (1985). Dynamical parameters of desorbing molecules. *Surf. Sci. Rep.*, 5:145 – 198.

Cremer, P., Stauners, C., Niemantsverdriet, J., Shen, Y., and Somorjai, G. (1995). The conversion of di-d bonded ethylene to ethylidyne of Pt(111) monitored with sum frequency generation: Evidence for an ethylidene (or ethyl) intermediate. *Surf.Sci.*, 328:111 – 118.

Daum, W., Krause, H.-J., Reichel, U., and Ibach, H. (1993). Identification of strained silicon layers at Si-SiO$_2$ interfaces and clean Si surfaces by nonlinear optical spectroscopy. *Phys. Rev. Lett.*, 71:1234 – 1237.

de Mul, M. and Mann, J. (1998). Determination of the thickness and optical properties of a langmuir film from the domain morphology by Brewster angle microscopy. *Langmuir*, 14:2455 – 2466.

Dewitz, J., Hübner, W., and Bennemann, K. (1996). Theory for nonlinear Mie-scattering from spherical metal clusters. *Z. Phys. D*, 37:75 – 84.

Domen, K., Fujino, T., Hirose, C., and Kano, S. (1998). Direct observation of short-lived unstable surface species by tunable picosecond infrared pulses. *Appl.Surf.Sci.*, 121:484 – 487.

Drude, P. (1891). Ueber die Reflexion und Brechung ebener Lichtwellen beim Durchgang durch eine mit Oberflächenschichten behaftete planparallele Platte. *Ann. der Physik*, 43:126 – 157.

Ducloy, M. and Fichet, M. (1991). General theory of frequency modulated selective reflection. Influence of atom surface interactions. *J. Phys. II France*, 1:1429 – 1446.

Erland, J., Bozhevolnyi, S., Pedersen, K., Jensen, J., and Hvam, J. (2000). Second-harmonic imaging of semiconductor quantum dots. *Appl. Phys. Lett.*, 77:806 – 808.

Erley, G. and Daum, W. (1998). Silicon interband transitions observed at Si(100)-SiO$_2$ interfaces. *Phys.Rev.B*, 58:R1734.

Euceda, A., Bylander, D., and Kleinman, L. (1983). Self-consistent electronic structure of 6- and 18-layer Cu(111) films. *Phys. Rev. B*, 28:528 – 534.

Failache, H., Saltiel, S., Fichet, M., Bloch, D., and Ducloy, M. (1999). Resonant van der Waals repulsion between excited Cs atoms and sapphire surface. *Phys. Rev. Lett.*, 83:5467 – 5470.

Feder, R., editor (1985). *Polarized Electrons in Surface Physics*. World Scientific, Singapore.

Feibelman, P. (1982). Surface electromagnetic fields. *Prog. Surf. Sci.*, 12:287 – 408.

Feinleib, J. (1966). Electroreflectance in metals. *Phys. Rev. Lett.*, 16:1200 – 1202.

Fichet, M., Schuller, F., Bloch, D., and Ducloy, M. (1995). van der Waals interactions between excited-state atoms and dispersive dielectric surfaces. *Phys. Rev. A*, 51:1553 – 1564.

Fischer, U. and Pohl, D. (1989). Observation of single-particle plasmons by near-field optical microscopy. *Phys.Rev.Lett.*, 62:458 – 461.

Fisher, R., editor (1983). *Optical Phase Conjugation*. Academic Press, New York.

Floersheimer, M. (1999). Second harmonic microscopy – a new tool for the remote sensing of interfaces. *Phys. Stat. Sol. (a)*, 173:15 – 27.

Floersheimer, M., Boesch, M., Brillert, C., Wierschem, M., and Fuchs, H. (1997). Second harmonic microscopy – a quantitative probe for molecular surface order. *Adv.Mat.*, 9:1061 – 1065.

Frederick, B., Cole, R., Power, J., Perry, C., Chen, Q., Richardson, N., Weightman, P., Verdozzi, C., Jennison, D., Schultz, P., and Sears, M. (1998). Molecular orientation with visible light: Reflectance-anisotropy spectroscopy of 3-thiophene carboxylate on Cu(110) surfaces. *Phys. Rev. B*, 58:10883 – 10889.

Fuchs, H. (1994). SXM-Methoden - nützliche Werkzeuge für die Praxis (in german). *Phys. Bl.*, 50:837–844.

Furtak, T. (1994). Electrochemical surface science. *Surf. Sci.*, 299/300:945 – 955.

Garcia-Vidal, F. and Pendry, J. (1996). Collective theory for surface enhanced Raman scattering. *Phys. Rev. Lett.*, 77:1163 – 1166.

Germer, T., Stephenson, J., Heilweil, E., and Cavanagh, R. (1993). Picosecond measurement of substrate-to-adsorbate energy transfer: The frustrated translation of CO/Pt(111). *J. Chem. Phys.*, 98:9986 – 9994.

Goodman, F. and Wachman, H. (1976). *Dynamics of Gas-Surface Scattering*. Academic Press, New York.

Gordon, J. and Swalen, J. (1977). The effect of thin organic films on the surface plasma resonance on gold. *Opt. Commun.*, 22:374 – 376.

Grafström, S. and Suter, D. (1996). Interaction of spin-polarized atoms with a surface studied by optical-reflection spectroscopy. *Phys. Rev. A*, 54:2169 – 2179.

Guo, J., Cooper, J., Gallagher, A., and Lewenstein, M. (1994). Theory of selective reflection spectroscopy. *Opt. Commun.*, 110:732 – 743.

Hache, F., Ricard, D., and Flytzanis, C. (1986). Optical nonlinearities of small metal particles: surface-mediated resonance and quantum size effects. *J. Opt. Soc. Am. B*, 3:1647 – 1655.

Hamermesh, A. (1962). *Group theory*. Addison Wesley, New York.

Hartschuh, A., Sanchez, E., Xie, X., and Novotny, L. (2003). High-resolution near-field raman microscopy of single-walled carbon nanotubes. *Phys.Rev.Lett.*, 90:095503–1 – 4.

Hecht, B., Bielefeldt, H., Novotny, L., Inouye, Y., and Pohl, D. (1996). Local excitation, scattering and interference of surface plasmons. *Phys.Rev.Lett.*, 77:1889 – 1892.

Hecht, B., Sick, B., Wild, U., Deckert, V., Zenobi, R., Martin, O., and Pohl, D. (2000). Scanning near-field optical microscopy with aperture probes: Fundamentals and applications. *J. Chem. Phys.*, 112:7761 – 7774.

Heilweil, E., Casassa, M., Cavanagh, R., and Stephenson, J. (1984). Picosecond vibrational energy relaxation of surface hydroxyl groups on colloidal silica. *J. Chem. Phys.*, 81:2856 – 2858.

Heilweil, E., Casassa, M., Cavanagh, R., and Stephenson, J. (1985). Vibrational deactivation of surface OH chemisorbed on SiO_2: Solvent effects. *J. Chem. Phys.*, 82:5216 – 5231.

Heinz, T. (1991). Second-order nonlinear optical effects at surfaces and interfaces. In Ponath, H.-E. and Stegeman, G., editors, *Nonlinear Surface Electromagnetic Phenomena*, pages 353 – 416. Elsevier.

Heinz, T., Chen, C., Ricard, D., and Shen, Y. (1982). Spectroscopy of molecular monolayers by resonant second-harmonic generation. *Phys. Rev. Lett.*, 48:478 – 481.

Heinz, T., Loy, M., and Thompson, W. (1985). Study of Si(111) sufaces by optical second-harmonic generation: Reconstruction and surface phase transformation. *Phys. Rev. Lett.*, 54:63 – 66.

Hénon, S. and Meunier, J. (1991). Microscope at the Brewster angle: Direct observation of first-order phase transitions in monolayers. *Rev. Sci. Instrum.*, 62:936 – 939.

Hicks, T. and Atherton, P. (1997). *The NanoPositioning Book*. Queensgate Instruments Limited, Berkshire.

Hingerl, K., Aspnes, D., Kamiya, I., and Florez, L. (1993). Relationship among reflectance-difference spectroscopy, surface photoabsorption, and spectroellipsometry. *Appl. Phys. Lett.*, 63:885 – 887.

Hoenig, D. and Moebius, D. (1991). Direct visualization of monolayers at the air-water interface by Brewster angle microscopy. *J. Phys. Chem.*, 95:4590 – 4592.

Hohlfeld, J., Matthias, E., Knorren, R., and Bennemann, K. (1997). Nonequilibrium magnetization dynamics of nickel. *Phys.Rev.Lett.*, 78:4861 – 4864.

Hua, X. and Gersten, J. (1986). Theory of second-harmonic generation by small metal spheres. *Phys. Rev. B*, 33:3756 – 3764.

Hunt, J., Guyot-Sionnest, P., and Shen, Y. (1987). Observation of C-H stretch vibrations of monolayers of molecules by optical sum-frequency generation. *Chem. Phys. Lett.*, 133:189 – 192.

Jensen, H., Reinisch, R., and Coutaz, J. (1997). Hydrodynamic study of surface plasmon enhanced non-local second-harmonic generation. *Appl.Phys.B*, 64:57 – 63.

Jeon, D., Hashizume, T., Sakurai, T., and Willis, R. (1992). Structural and electronic properties of ordered single and multiple layers of Na on the Si(111) surface. *Phys.Rev.Lett.*, 69:1419 – 1422.

Joannopoulos, J., Meade, R., and Winn, J. (1995). *Photonic Crystals*. Princeton Press, Princeton N.J.

Jordan, C., Marowsky, G., and Rubahn, H.-G. (1995). Coverage dependent changes in the azimuthal anisotropy of second harmonic generated from Na/Si(111) 7×7 interfaces. *Opt. Comm.*, 120:98 – 102.

Jostell, U. (1979). Plasmons in monolayer Na, K and Rb films. *Surf.Sci.*, 82:333 – 348.

Kawata, S., Inouye, Y., and Sugiura, T. (1994). Near-field scanning optical microscope with a laser trapped probe. *Jpn. J. Appl. Phys.*, 33:L1725 – L1727.

Kelly, P., Tang, Z.-R., Woolf, D., Williams, R., and McGilp, J. (1991). Optical second harmonic generation from Si(111)-As and Si(100)-As. *Surf. Sci.*, 251/252:87 – 91.

Kim, J., Kim, J., Song, K.-B., Lee, S., Kim, E.-K., Choi, S.-E., Lee, Y., and Park, K.-H. (2003). Near-field imaging of surface plasmon on gold nano-dots fabricated by scanning probe lithography. *J.Microsc.*, 209:236 – 240.

Kino, G. and Chim, S. (1990). Mirau correlation microscope. *Appl.Opt.*, 29:3775 – 3783.

Kirilyuk, A., Petukhov, A., Rasing, T., Megy, R., and Beauvillain, P. (1997a). Second harmonic generation study of quantum well states in thin noble metal overlayer films. *Surf.Sci.*, 377:409 – 413.

Kirilyuk, V., Kirilyuk, A., and Rasing, T. (1997b). A combined nonlinear and linear magneto-optical microscopy. *Appl.Phys.Lett.*, 70:2306 – 2308.

Kirilyuk, V., Kirilyuk, A., and Rasing, T. (1997c). New mode of domain imaging: Second harmonic generation microscopy. *J.Appl.Phys.*, 81:5014.

Klar, T. and Hell, S. (1999). Subdiffraction resolution in far-field fluorescence microscopy. *Opt. Lett.*, 24:954 – 956.

Knoll, B. and Kellmann, F. (1999). Near-field probing of vibrational absorption for chemical microscopy. *Nature*, 399:134 – 137.

Kolb, D. (1982). The study of solid-liquid interfaces by surface plasmon polariton excitation. In Agranovich, V. and Mills, D., editors, *Surface Polaritons*, pages 299 – 329. North-Holland Publishing Company.

Koopmans, B., F., der Woude, and Sawatzky, G. (1992). Surface symmetry resolution of nonlinear optical techniques. *Phys. Rev. B*, 46:12780 – 12783.

Koopmans, B., Janner, A.-M., Jonkman, H., Sawatzky, G., and van der Woude, F. (1993). Strong bulk magnetic dipole induced second-harmonic generation from C_{60}. *Phys. Rev. Lett.*, 71:3569 – 3572.

Koopmans, B., Koerkamp, M., Rasing, T., and Berg, H. V. D. (1995). Observation of large Kerr angles in the nonlinear optical response from magnetic multilayers. *Phys.Rev.Lett.*, 74:3692 – 3695.

Kötz, R., Kolb, D., and Sass, J. (1977). Electron density effects in surface plasmon excitation on silver and gold electrodes. *Surf. Sci.*, 69:359 – 364.

Kreibig, U. (1997). Optics of nanosized metals. In Hummel, R. and Wißmann, P., editors, *Handbook of Optical Properties, Vol.II, Optics of Small Particles, Interfaces, and Surfaces*, pages 145 – 190. CRC Press, Boca Raton.

Kretschmann, E. and Raether, H. (1968). Radiative decay of non-radiative surface plasmons excited by light. *Z. Naturforschung*, 23a:2135 – 2136.

Landau, L. and Lifshitz, E. (1963). *Electrodynamics of Continuous Media*. Pergamon Press, Oxford.

Landau, L. and Lifshitz, E. (1978). *Mechanics*. Pergamon Press, Oxford.

Landau, L. and Lifshitz, E. (1980). *The Classical Theory of Fields*. Pergamon Press, Oxford.

Landau, L. and Lifshitz, E. (1988). *Quantum Mechanics: Non-Relativistic Theory*. Pergamon Press, New York.

Lee, Y.-S., Anderson, M., and Downer, M. (1997). Fourth-harmonic generation at a crystalline GaAs(001) surface. *Opt.Lett.*, 22:973 – 975.

Letokhov, V. (1975). Possible laser modification of field ion microscopy. *Phys. Lett.*, 51A:231 – 232.

Leung, P. and George, T. (1989). Molecular spectroscopy at corrugated metal surfaces. *Spectroscopy*, 4:35 – 41.

Liebsch, A. (1989). Second-harmonic generation from alkali-metal overlayers. *Phys. Rev. B*, 40:3421 – 3424.

Lifshitz, E. and Pitaevskii, L. (1986). *Physical Kinetics*. Pergamon Press, Oxford.

Lüpke, G., Bottomley, D., and van Driel, H. (1994). Resonant second-harmonic generation on Cu(111) by a surface-state to image-potential-state transition. *Phys. Rev. B*, 49:17303 – 17306.

Magonov, S. and Whangbo, M.-H. (1996). *Surface Analysis with STM and AFM*. VCH, Weinheim.

Mandel, L. and Wolf, E. (1995). *Optical Coherence and Quantum Optics*. Cambridge Universtiy Press, New York.

Maradudin, A. and Mills, D. (1975). Scattering and absorption of electromagnetic radiation by a semi-infinite medium in the presence of surface roughness. *Phys. Rev. B*, 11:1392 – 1415.

Marowsky, G., Chi, L., Möbius, D., Steinhoff, R., Shen, Y., Dorsch, D., and Rieger, B. (1988). Non-linear optical properties of hemicyanine monolayers and the protonation effect. *Chem.Phys.Lett.*, 147:420–424.

McGilp, J. (1987). Determining metal-semiconductor interface structure by optical second-harmonic-generation. *J. Vac. Sci. Technol. A*, 5:1442 – 1446.

McIntyre, J. and Aspnes, D. (1971). Differential reflection spectroscopy of very thin surface films. *Surf. Sci.*, 24:417 – 434.

Michaelis, J., Hettich, C., Mlynek, J., and Sandoghdar, V. (2000). Optical microscopy using a single-molecule light source. *Nature*, 405:325 – 328.

Mirlin, D. (1982). Surface phonon polaritons in dielectrics and semiconductors. In Agranovich, V. and Mills, D., editors, *Surface Polaritons*, pages 3 – 67. North-Holland Publishing Company.

Mizrahi, V. and Sipe, J. (1988). Phenomenological treatment of surface second-harmonic generation. *J. Opt. Soc. Am. B*, 5:660 – 667.

Monk, P. (2001). *Fundamentals of Electroanalytical Chemistry*. John Wiley & Sons, Chichester.

Morin, M., Jakob, P., Levinos, N., Chabal, Y., and Harris, A. (1992). Vibrational energy transfer on hydrogen-terminated vicinal Si(111) surfaces: Interadsorbate energy flow. *J. Chem. Phys.*, 96:6203 – 6212.

Morris, V., Gunning, A., and Kirby, A. (1999). *Atomic Force Microscopy for Biologists*. Imperial College Press, London.

Müller, T., Vaccaro, P., Balzer, F., and Rubahn, H.-G. (1997). Size dependent optical second harmonic generation from surface bound Na clusters: Comparison between experiment and theory. *Opt. Comm.*, 135:103 – 108.

Murphy, R., Yeganeh, M., Song, K., and Plummer, E. (1989). Second-harmonic generation from the surface of a simple metal, Al. *Phys. Rev. Lett.*, 63:318 – 321.

Nachet, C. (1847). *Compte Rendu de l'Academie des Sciences*, XXIV:976.

Nienhuis, G., Schuller, F., and Ducloy, M. (1988). Nonlinear selective reflection from an atomic vapor at arbitrary incidence angle. *Phys. Rev. A*, 38:5197 – 5205.

Oria, M., Chevrollier, M., Bloch, D., Fichet, M., and Ducloy, M. (1991). Spectral observation of surface-induced van der waals attraction on atomic vapour. *Europhys. Lett.*, 14:527 – 532.

Östling, D., Stampfli, P., and Bennemann, K. (1993). Theory of nonlinear optical properties of small metallic spheres. *Z. Phys. D*, 28:169 – 175.

Otto, A. (1968). Wechselwirkung elektromagnetischer Oberflächenwellen. *Z. Angew. Physik*, 27:207 – 209.

Paesler, M. and Mojer, P. (1996). *Near Field Optics*. John Wiley, New York.

Pan, R.-P., Wei, H., and Shen, Y. (1989). Optical second-harmonic generation from magnetized surfaces. *Phys. Rev. B*, 39:1229 – 1234.

Papageorgopoulos, C. and Kamaratos, M. (1992). The behaviour of Na on 1x1 and 7x7 structures of Si(111) and its effect on the oxidation of these structures. *J.Phys.: Condens.Matter*, 4:1935 – 1945.

Pohl, D. (1982). Optical near-field microscope. European Patent Application No.0112401.

Pohl, D. and Courjon, D. (1993). *Near Field Optics*. Kluwer, Dordrecht.

Pohl, D., Denk, W., and Lanz, M. (1984). Optical stethoscopy: Image recording with resolution $\lambda/20$. *Appl. Phys. Lett.*, 44:651 – 653.

Prieve, D. and Walz, J. (1993). Scattering of an evanescent surface wave by a microscopic dielectric sphere. *Appl. Opt.*, 32:1629 – 1641.

Qian, J.-P. and Wang, G.-C. (1990). A simple ultrahigh vacuum surface magneto-optical Kerr effect setup for the study of surface magnetic anisotropy. *J. Vac. Sci. Technol. A*, 8:4117 – 4119.

Rabi, O., Amy-Klein, A., Saltiel, S., and Ducloy, M. (1994). Three-level non-linear selective reflection at a dielectric/Cs vapour interface. *Europhys. Lett.*, 25:579 – 585.

Raether, H. (1988). *Surface Plasmons on Smooth and Rough Surfaces and on Gratings*. Springer-Verlag, Berlin.

Rayleigh, L. (1879). Investigations in optics with special reference to the spectroscope. *Philos. Mag.*, 8:261 – 274.

Reddick, R., Warmack, R., and Ferrell, T. L. (1989). New form of scanning optical microscopy. *Phys. Rev. B*, 39:767 – 770.

Reif, J., Rau, C., and Matthias, E. (1993). Influence of magnetism on second harmonic generation. *Phys. Rev. Lett.*, 71:1931 – 1934.

Reif, J., Zink, J., Schneider, C., and Kirschner, J. (1991). Effects of surface magnetism on optical second harmonic generation. *Phys. Rev. Lett.*, 67:2878 – 2881.

Richardson, N. and Bradshaw, A. (1979). The frequencies and amplitudes of CO vibrations at a metal surface from model cluster calculations. *Surf. Sci.*, 88:255 – 268.

Rubahn, H.-G. (1999). *Laser Applications in Surface Science and Technology*. John Wiley & Sons Limited.

Rubahn, H.-G. (2004). *Nanophysik und Nanotechnologie*. B.G. Teubner Verlag, Wiesbaden.

Sanchez, E., Novotny, L., and Xie, X. (1999). Near-field fluorescence microscopy based on two-photon excitation with metal tips. *Phys.Rev.Lett.*, 82:4014 – 4017.

Sandroff, C., Turco-Sandroff, F., Florez, L., and Harbison, J. (1991). Recombination at GaAs surfaces and GaAs/AlGaAs interfaces probed by *in situ* photoluminescence. *J. Appl. Phys.*, 70:3632 – 3635.

Sarid, D. (1997). *Scanning Force Microscopy: With Applications to Electric, Magnetic and Atomic Forces*. Oxford University Press, Oxford.

Sautenkov, V., Velichanskii, V., Zibrov, A., Luk'yanov, V., Nikitin, V., and Tyurikov, D. (1981). Intra-Doppler resonances of the cesium D_2 line in a selective specular reflection profile. *Kvantovaya Electronika*, 8:1867 – 1872.

Scholl, A., Baumgarten, L., Jacquemin, R., and Eberhardt, W. (1997). Ultrafast spn dynamics of ferromagnetic thin films observed by f-sec spin resolved two photon photoemission. *Phys.Rev.Lett.*, 79:5146 – 5149.

Schönflies, A. (1891). *Kristallsysteme und Kristallstruktur*. Leipzig.

Schrader, M., Hell, S., and van der Voort, H. (1998). Three-dimensional super-resolution with a 4π-confocal microscope using image restoration. *J. Appl. Phys.*, 84:4033 – 4042.

Schuurmans, M. (1976). Spectral narrowing of selective reflection. *J. Phys. (Paris)*, 37:469 – 485.

Sekatskii, S. and Letokhov, V. (1996). Scanning optical microscopy with a nanometer spatial resolution based on resonant excitation of fluorescence from one-atom excited center. *JETP Letters*, 63:319–323.

Shen, Y. (1984). *The Principles of Nonlinear Optics*. John Wiley, New York.

Shen, Y. (1986). Surface second harmonic generation. *Ann. Rev. Mater. Sci.*, 16:69 – 86.

Sheppard, C. (1996). Image formation in three-photon fluorescence microscopy. *Bioimaging*, 4:124 – 128.

Shockley, W. (1939). On the surface states associated with a periodic potential. *Phys. Rev.*, 56:317 – 323.

Shubeita, G., Sekatskii, S., Chergui, M., Dietler, G., and Letokhov, V. (1999). Investigation of nanolocal fluorescence resonance energy transfer for scanning probe microscopy. *Appl. Phys. Lett.*, 74:3453 – 3455.

Simon, H., Mitchell, D., and Watson, J. (1975). Second harmonic generation with surface plasmons in alkali metals. *Opt. Comm.*, 13:294 – 298.

Sipe, J., Moss, D., and van Driel, H. (1987). Phenomenological theory of optical second- and third-harmonic generation from cubic centrosymmetric crystals. *Phys. Rev. B*, 35:1129 – 1141.

Sivukhin, D. (1948). Molekulyarnaya teoriya otrazheniya i prelomleniya sveta. *Zh. Eksperim. Teor. Fiz. SSSR*, 18:976 – 994.

Sivukhin, D. (1951). Ellipticheskaya polyarizatziya pri otrazhenii sveta ot zhidkostej. *Zh. Eksperim. Teor. Fiz. SSSR*, 21:367 – 376.

Smilowitz, L., Jia, Q., Yang, X., Li, D., McBranch, D., Buelow, S., and Robinson, J. (1997). Imaging nanometer-thick patterned self-assembled monolayers via second-harmonic generation microscopy. *J.Appl.Phys.*, 81:2051 – 2054.

Sommerfeld, A. (1909). Über die Ausbreitung der Wellen in der drahtlosen Telegraphie. *Ann. Phys. Leipz.*, 28:665 – 737.

Song, K., Heskett, D., Dai, H.-L., Liebsch, A., and Plummer, E. (1988). Dynamical screening at a metal surface probed by second-harmonic generation. *Phys. Rev. Lett.*, 61:1380 – 1383.

Soukissian, P., Bakshi, M., Hurych, Z., and Gentle, T. (1989). Electronic properties of alkali metal/silicon interfaces: A new picture. *Surf.Sci.*, 221:L759 – L768.

Specht, M., Pedarnig, J., Heckl, W., and Hänsch, T. (1992). Scanning plasmon near-field microscope. *Physical Review Letters*, 68:476 – 479.

Spierings, G., Koutsos, V., Wierenga, H., Prins, M., Abraham, D., and Rasing, T. (1993). Optical second harmonic generation study of interface magnetism. *Surface Science*, 287/288:747 – 749.

Stegemann, G., Fortenberry, R., Karaguleff, C., Moshrefzadeh, R., III, W. H., Wyck, N. V., and Sipe, J. (1983). Coherent anti-Stokes Raman scattering in thin-film dielectric waveguides. *Opt. Lett.*, 8:295 – 297.

Stratton, J. (1941). *Electromagnetic Theory*. McGraw-Hill Book Company, New York.

Sturm, H., Balzer, F., and Rubahn, H.-G. (2003). Bundesanstalt für Materialwissenschaften Berlin. unpublished.

Synge, E. (1928). Suggested method for extending microscopic resolution into the ultra-microscopic region. *Phil. Mag.*, 6:356 – 362.

Tamm, I. (1932). Über eine mögliche Art der Elektronenbindung an Kristalloberflächen. *Z. Phys.*, 76:849 – 850.

Temple, P. (1981). Total internal reflection microscopy: A surface inspection technique. *Appl. Opt.*, 20:2656 – 2664.

Todd, B. and Eppell, S. (2001). A method to improve the quantitative analysis of SFM images at the nanoscale. *Surf. Sci.*, 491:473 – 483.

Tom, H., Mate, C., Zhu, X., Crowell, J., Shen, Y., and Somorjai, G. (1986). Studies of alkali adsorption on Rh(111) using optical second-harmonic generation. *Surface Science*, 172:466 – 476.

Tsang, T. (1995). Optical third-harmonic generation at interfaces. *Phys. Rev. A*, 52:4116 – 4125.

Tsankov, D., Hinrichs, K., Röseler, A., and Korte, E. (2001). FTIR ellipsometry as a tool for studying organic layers: From Langmuir-Blodgett films to can coatings. *phys. stat. sol. (a)*, 188:1319 – 1329.

Tüshaus, M., Schweizer, E., Hollins, P., and Bradshaw, A. (1987). Yet another vibrational study of the adsorption system Pt{111}-CO. *J. Electron Spectrosc. Relat. Phenom.*, 44:305 – 316.

Urbach, L., Percival, K., Hicks, J., Plummer, E., and Dai, H.-L. (1992). Resonant surface second-harmonic generation: Surface states on Ag(110). *Phys. Rev. B*, 45:3769 – 3772.

Vartanyan, T. (1985). Resonant reflection of intense optical radiation from a low-density gaseous medium. *Sov. Phys. JETP*, 61:674 – 677.

Verbiest, T., Elshicht, S. V., Kauranen, M., Hellemans, L., Snauwaert, J., Nuckolls, C., Katz, T., and Persoons, A. (1998). Strong enhancement of nonlinear optical properties through supramolecular chirality. *Science*, 282:913 – 915.

Vohnsen, B., Bozhevolnyi, S., Pedersen, K., Erland, J., Jensen, J., and Hvam, J. (2001). Second-harmonic scanning optical microscopy of semiconductor quantum dots. *Opt. Commun.*, 189:305 – 311.

Volkov, V., Bozhevolnyi, S. I., Bordo, V., and Rubahn, H.-G. (2004). Near field imaging of organic nanofibres. *J. Microscopy*, 215:241 – 244.

Wang, C. and Duminski, A. (1968). Second-harmonic generation of light at the boundary of alkali halides and glasses. *Phys. Rev. Lett.*, 20:668 – 671.

Weightman, P. (2001). The potential of reflection anisotropy spectroscopy as a probe of molecular assembly on metal surfaces. *phys. stat. sol. (a)*, 188:1443 – 1453.

Weis, A., Sautenkov, V., and Hänsch, T. (1992). Observation of ground-state Zeeman coherences in the selective reflection from cesium vapor. *Phys. Rev. A*, 45:7991 – 7996.

Wierenga, H., Jong, W. D., Prins, M., Rasing, T., Vollmer, R., Kirilyuk, A., Schwabe, H., and Kirschner, J. (1995). Interface magnetism and possible quantum well oscillations in ultrathin Co/Cu films observed by magnetization induced second harmonic generation. *Phys.Rev.Lett.*, 74:1462 – 1465.

Wiesendanger, R. (1994). *Scanning Probe Microscopy and Spectroscopy: Methods and Applications*. Cambridge University Press, Cambridge.

Wiesendanger, R. (1998). *Scanning Probe Microscopy: Analytical Methods (Nanoscience and Technology)*. Springer, Berlin.

Wijekoon, W., Ho, Z., and Hetherington, W. (1987). Ethylene adsorption on ZnO: CARS spectroscopy with optical waveguides. *J. Chem. Phys.*, 86:4384 – 4390.

Wilson, T. (1990). *Confocal Microscopy*. Academic Press, London.

Woerdman, J. and Schuurmans, M. (1975). Spectral narrowing of selective reflection from sodium vapour. *Opt. Commun.*, 14:248 – 251.

Wokaun, A., Bergman, J., Heritage, J., Glass, A., Liao, P., and Olson, D. (1981). Surface second-harmonic generation from metal island films and microlithographic structures. *Phys. Rev. B*, 24:849 – 856.

Wood, R. (1909). Selective reflection of monochromatic light by mercury vapour. *Phil. Mag.*, 18:187 – 193.

Wu, Q., Grober, R., Gammon, D., and Katzer, D. (2000). Excitons, biexcitons, and electron-hole plasma in a narrow 2.8-nm $GaAs/Al_xGa_{1-x}As$ quantum well. *Phys. Rev. B*, 62:13022 – 13027.

Wyant, J. (1995). Computerized interferometric measurement of surface microstructure. *Proc.SPIE*, 2576:122–130.

Wylie, J. and Sipe, J. (1984). Quantum electrodynamics near an interface. *Phys. Rev. A*, 30:1185 – 1193.

Wylie, J. and Sipe, J. (1985). Quantum electrodynamics near an interface. II. *Phys. Rev. A*, 32:2030 – 2043.

Yariv, A. (1985). *Optical Electronics*. Holt-Saunders, New York.

Yariv, A. (1989). *Quantum Electronics*. John Wiley & Sons, New York, 3rd edition.

Ye, P. and Shen, Y. (1983). Local-field effect on linear and nonlinear optical properties of adsorbed molecules. *Phys.Rev.B*, 28:4288 – 4294.

Yee, H., Byers, J., and Hicks, J. (1994). A nonlinear optical study of chiral surfaces. *Proc.SPIE*, 2125:119 – 131.

Ying, Z., Wang, J., Andronica, G., Yao, J.-Q., and Plummer, E. (1993). Azimuthal and incident angle dependences in the second-harmonic generation from aluminum. *J. Vac. Sci. Technol. A*, 11:2255 – 2259.

Zangwill, A. (1988). *Physics at Surfaces*. Cambridge University Press, Cambridge.

Zenhausern, F., O'Boyle, M., and Wickramasinghe, H. (1994). Apertureless near-field optical microscope. *Appl. Phys. Lett.*, 65:1623 – 1625.

Zernicke, F. (1953). Nobel prize lecture. *The Nobel Foundation*.

Zhdanov, V. and Zamaraev, K. (1982). Vibrational relaxation of adsorbed molecules. Mechanism and manifestation in chemical reactions on solid surface. *Catal. Rev. Sci. Eng.*, 24:373 – 413.

Zhizhin, G., Morozov, N., Moskalova, M., Sigarov, A., Shomina, E., Yakovlev, V., and Grigos, V. (1980). Surface electromagnetic wave absorption on copper surfaces with Langmuir films using CO_2 laser excitation. *Thin Solid Films*, 70:163 – 168.

Zhizhin, G., Moskalova, M., Shomina, E., and Yakovlev, V. (1982). Surface electromagnetic wave propagation on metal surfaces. In Agranovich, V. and Mills, D., editors, *Surface Polaritons*, pages 93 – 144. North-Holland Publishing Company.

Zhu, X., Shen, Y., and Carr, R. (1985). Correlation between thermal desorption spectroscopy and optical second harmonic generation for monolayer surface coverages. *Surf. Sci.*, 163:114 – 120.

Zink, J., Reif, J., and Matthias, E. (1992). Water adsorption on (111) surfaces of BaF_2 and CaF_2. *Phys.Rev.Lett.*, 68:3595 – 3598.

Index

Optics and Spectroscopy at Surfaces and Interfaces. Vladimir G. Bordo and Horst-Günter Rubahn
Copyright © 2005 WILEY-VCH Verlag GmbH & Co. KGaA, Weinheim
ISBN: 3-527-40560-7